I0414649

A Coal Miner's Family at Mooseheart

By

Ernest L. Rhodes

authorHOUSE®

AuthorHouse™
1663 Liberty Drive
Bloomington, IN 47403
www.authorhouse.com
Phone: 1-800-839-8640

© 2010 Ernest L. Rhodes. All rights reserved.

No part of this book may be reproduced, stored in a retrieval system, or
transmitted by any means without the written permission of the author.

First published by AuthorHouse 5/11/2010

ISBN: 978-1-4520-0950-6 (sc)
ISBN: 978-1-4520-0951-3 (e)

Library of Congress Control Number: 2010904744

Printed in the United States of America
Bloomington, Indiana

This book is printed on acid-free paper.

This work is dedicated to James J. Davis,
the founder of Mooseheart
and
to every member of the Moose fraternity,
past and present, woman and man,
the heart and soul of Mooseheart

TABLE OF CONTENTS

INTRODUCTION

A COAL MINER'S FAMILY at *Mooseheart* describes the lives of Homer Rhodes' widow and their children at Mooseheart, IL, from 1919-1939.

Arguably the most unusual child care program in the 20th Century, the orphans' home-school, Mooseheart, was founded in 1913 by James J. Davis, and the Moose Fraternity. Davis was an idealistic visionary who assembled a group of compassionate experts to create and operate this city of children.

Ours is the story of my family. It begins with our arrival at Mooseheart and my rough introduction to the boys' codes of behavior enforced by peer pressure. Then I report on how a Demerit system was replaced by a Merit system that worked well for 1,300 students living in this community which absolutely prohibited corporal punishment. Also I explain our daily routines.

Further sections illustrate how Davis's idealistic vision worked for our family and the Mooseheart students we knew: 1) how Blanche and her family adopt Earl and Carolyn Guinn, who have lost both parents, 2) how students play and compete as the Mooseheart Spirit emerges, 3) how they can work to earn and spend their own money, 4) how they must learn a skilled trade, 5) how they may get a high school diploma—if they can pass the courses, 6) how they worship in the faith of their parents, 7) how they dance and romance, 8) how they dream, strive, become lonesome, suffer growing pains, 9) how they become ambitious, develop enough courage to leave Mooseheart to scatter and settle; 10) And finally, how they return to their very special utopia and wonder whether they can ever repay the Moose.

Since this is a family memoir, I attach sections about our heritage and our life before Mooseheart. I record what we know about our parents, Blanche Porter and, Homer Rhodes, and about Carrie Thomas, our mother's birth mother. I attach also a section about Spruce Knob, WV, and the Elk Lick Coal Company when Homer Rhodes was Superintendent of the mine there from 1919 until he died in 1925. With the help of my sons Stanley and Lloyd Rhodes and Joe Postek, my childhood playmate, I have reconstructed the Spruce Knob community and the mining done there. My enduring memories are augmented by old photographs and clippings, as well as many returns visits to Fork Mountain.

The incidents that I record in this story are true, as are most of the names, as far as I recall. My girl friends in the section on "Dancing and Romancing" are composites given tag names to convey truth rather than accuracy.

Let me thank those who have enabled me to assemble and prepare photographs, some going back almost 100 years. My sister Dorothy Guinn saved the pictures and supplied information about them. Arthur Daniels, her son-in-law, resized and repaired all of the pictures. Steve Watson, my computer Guru, provided technical help in reproducing them and often coaxed my laptop to work.

Richard Meyers helped me understand such technicalities as DPI'S.

Babette Meyers first suggested that I should write and share the stories I told of Mooseheart. I have had dozens of readers and am particularly indebted to Dr. Janet Bing and her husband the late Dr. Charles Ruhl, to Dr. Nancy Bazin for her severe critiques and many concrete suggestions. Gloria Wilson and Dr. Jack Wilson have also stood out as encouraging and perceptive readers.

Almost three decades of stories and information exchanged at the October Homecomings at Mooseheart enliven this work; I am grateful to Charlotte Rapp who held our Mooseheart classmates together with her *Rapp Sheet* and gatherings of us "Golden Oldies" until we began to die off in recent years.

As I noted, my sons Stanley and Lloyd Rhodes along with Joe Postek, joined me in exploring the sites of the Spruce Knob community and traces of the Elk Lick Coal Company mine in the hills of West Virginia.

I offer my very special thanks to Karen Vaughan, Digital Services Coordinator at the Old Dominion University's Perry Library, her expertise and calm judgment, for teaching me to solve dozens of my technical problems.

My wife and my best friend Dr. Carolyn Hodgson Meyers Rhodes tirelessly helped with the heavy lifting during the many revisions of this book.

This book began with family records. The earliest versions, titled Dorothy's Album, featured documentation added to many scrapbooks created by Dorothy Rhodes Guinn. All that she and her husband, the late Dr. Earl Guinn, Mooseheart, 1932, gathered about the

Rhodes heritage melded inextricably with our memories of decades at Mooseheart. Our Mother's *Journal* and *Little Book* added yet another dimension. So did public documents and the recollections of classmates

Gradually, I realized that very few lengthy personal testimonies about the Mooseheart experience have been published. So I hope that our family's story will convey the particular quality and depth of our years in the City of Children.

I believe I speak for many of the 11,000 or more former Mooseheart students. Certainly, for those I have personally known, I can affirm that we owe the Moose Fraternity a debt we can never fully repay. We proudly acknowledge our gratitude.

ARRIVING AT MOOSEHEART

WHEN OUR FATHER died on March 9, 1925 he left very little to support our mother and their family except his membership in the Moose fraternity. It allowed mother to apply for admission to Mooseheart, the Orphans' home-school that the Moose established in Illinois in 1913 to care for the families of members who died. With the assistance of the Richwood Moose lodge, mother applied for admission to Mooseheart and our family was accepted.

My sister Dorothy Guinn and I are the two Rhodes children still alive. I was 11 years old and she was 8 when we entered Mooseheart on July 26, 1926. She says that we left Richwood on July 23, which marked the end of our life together as a family under a single roof. But that fact mattered very little because we were leaving Carrie Thomas Meyers, our mother's birth mother, and Daddy George Meyers, with whom we had been living after our father died. To this day, Dorothy and I agree they were an unpleasant pair who didn't understand us or like us and who made our life miserable. Later our move to a new world would matter even more as we began to realize that we were becoming a part of a large nurturing family in many warm rooms under solid roofs.

An article in the Richwood newspaper, I believe it was called the *Nicholas County Republican,* gathers the essential facts about our prospects:

> Mrs. Blanche Rhodes and children, Ernest, Clifford, Dorothy and Clyde left Sunday morning for Mooseheart, Ill., accompanied by G.P. White of the Richwood Lodge Loyal Order of Moose.
>
> These children will receive a high school education and a trade there, before going out in life for them-selves, while the mother will be employed in the service department of this great institution with compensation
>
> At the present time this great Mooseheart Home is educating and training fourteen hundred boys and girls of departed members of the Moose fraternity.
>
> We are proud to learn that they are fortunate in securing this valuable privilege of Mooseheart service. They made the trip by automobile.

The article in the Richwood paper was accompanied by a picture of our family with mother dressed in black (Photo 1)

The three day drive from Richwood to Mooseheart has left a few spots of time in my memory. We drove through Huntington, West Virginia, where the city blocks seemed

Photo 1—Dressed for Mooseheart.

regular and endless. We crossed a bridge into Ironton, Ohio. There our mother suffered another heart attack and was hospitalized. We traveled on with Mr. White, hoping she could join us in a few days. Vaguely, I recall that we spent the night in Portsmouth, Ohio. I remember vividly the rolling land of Ohio, so different from the choppy hills of my West Virginia. More monotonous city blocks and slow going characterized Indianapolis. Long after, when Eisenhower's Interstate Highway system by-passed the city, I thought back to our tedious hours there.

For Terre Haute I recall only the sign on the road; it puzzled me until I studied Latin in college. Paris, a tiny town just over the Ohio line into Illinois, drew my attention because I had heard of Paris in France. We spent the night in Danville, Illinois, where it was dark and starless, but not unpleasant or frightening.

We reached Mooseheart the next afternoon. Mr. White, who had driven us to Mooseheart from the Richwood Moose Lodge, turned us over to Mrs. York in the Campanile.

As we would learn, Mrs. York was the grandmother of two Mooseheart girls, Dorothy and Elizabeth York. A handsome white-headed motherly woman, Mrs. York was the official greeter of visitors for many years—a beloved person and a fixture in our lives as long as we were at Mooseheart. For a time, one of Blanche Rhodes' jobs was to help her and fill in for her on her days off

In retrospect, we could not then foresee the wondrous change Mooseheart would make in our family's life. What a contrast it offered: we crawled out of a grim pit at 16 Riverside Drive, Richwood, WV, to be dazzled by the vistas of widespread acres on the Illinois prairie that reached toward tomorrow.

James J. Davis, once an immigrant Welsh steelworker, gathered colleagues to found Mooseheart in 1913. During the rest of the 20th Century it would become the home for more than 11,000 children like us who had lost one or both parents.

Mooseheart children always bristled at the term "Orphans Home," to describe this remarkable place. We were not orphans. Mooseheart was our school and our home—for a time. We were not orphans! We had a family--our fathers' brothers—thousands of them nationwide who supported us with a portion of their meager wages. Their wages were meager because they were, most of them, like our natural fathers, low-paid workers in steel mills, sewers, coal mines, salesrooms, on farms and in other places where men are regularly underpaid for their skills and sweat. Our fathers' brothers called themselves Moose.

Some of us had mothers—so all of us had mothers: Mooseheart Mothers who shared their affection—their devotion with all of the children. (Photo 2) And we had each other: To this day former Mooseheart Kids are bound by ties similar those of blood brothers and sisters fortunate enough to have lived together in security and affection in a regular home.

After Mr. White turned us over to Mrs. York, she sent our sister Dorothy to Detention in Pine Cottage on the Girls Campus and assigned Clifford, Clyde and me to Detention in Walnut Cottage, near the Mooseheart Lake. All newly arrived kids were isolated, to prevent them from bringing in diseases and unwanted critters like nits and lice which could have caused serious problems for the general population. Mooseheart did its best to protect us from such things.

Mother had been released from the hospital in Ohio and reached Mooseheart on August 2. She spent several hours on her second day there going through the hair of one of the new kids with a fine tooth comb—picking out head lice. It wasn't one of her kids. She had checked us thoroughly before we left Richwood.

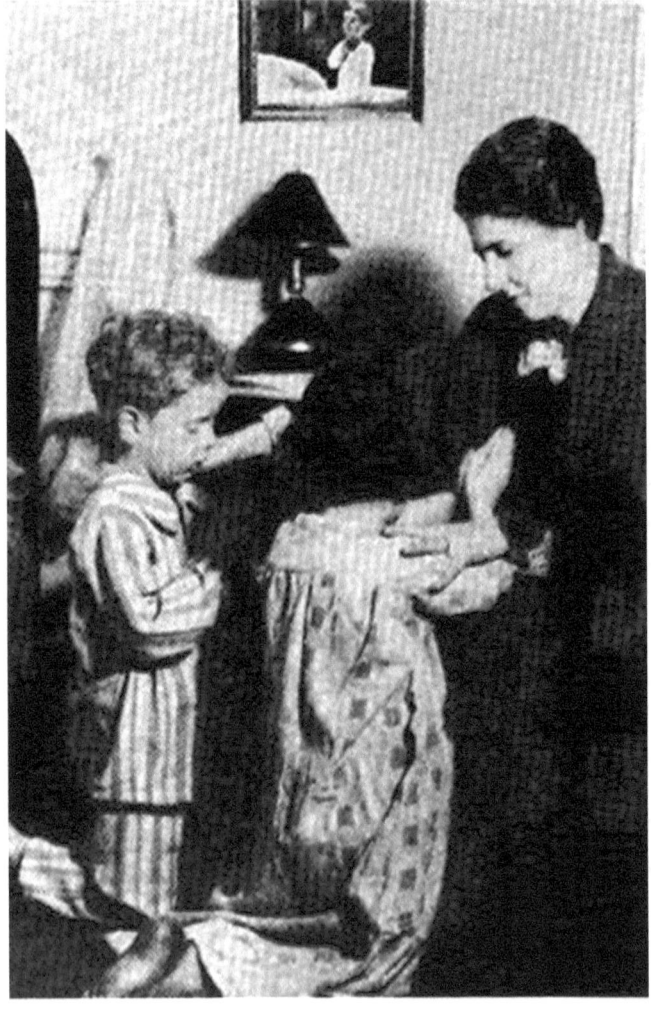

Photo 2 —Blanche an Iconic Mooseheart Mother.

We lived with Mrs. McLucas and seven other new guys in Walnut Cottage for two weeks of Detention. She had three boys and two girls of her own, but she was never assigned to take care of her own children. Instead her job was to take care of other Mooseheart Kids—including many who had no father or mother. Like most Mooseheart Mothers, she was a "nifty" or a "darby" person. In the language of the "Kids" at Mooseheart, she was a good person. If we had not liked her, we could have described her as a "lousy" or a "snotty" person.

Severe injury and death were rare at Mooseheart. I know of only four kids who died at Mooseheart. One of them, James Kuhn, died from an infection or blood poisoning before we were admitted. Two were drowned at different times while swimming without supervision—one in Mooseheart Lake and the other in the Fox River. The only student that I actually knew, George Duffield, died of pneumonia.

Dr. Johnny Nichols saw that we received almost every shot available for typhoid fever, scarlet fever, diphtheria, whooping cough, and chicken pox. Of course, we were vaccinated for smallpox.

We were subjected to tests designed to measure intelligence and to label our temperament and aptitudes. These tests were administered by the staff of Dr. Martin Luther Reymert, a psychologist, who directed the Child Research Laboratory, housed in a building funded by Moose Lodges of Ohio. We called the place "The Bug House."

While I was working as a Gofer in the Reference Room of the West Virginia University Library in 1935, I was asked to get an issue of a periodical, *School and Soci-*

ety. In one issue I noticed an article by Dr. Reymert about the effect of movies on the attitudes of children concerning war and crime. I am sorry to reveal the study may have been flawed. I was part of the Control Group that was NOT permitted to see *All Quiet on Western Front, Journey's End,* and *The Big House.* Even so, I sneaked into the movies with a forged ticket--and those films made a lasting impression upon my ideas about war and crime. I sometimes have a sense of guilt about messing up the good doctor's study. After all, he was the first published scholar I ever knew.

Dr. Reymert was a wonderful human as were Mooseheart Mothers, most of our teachers and some of the administrators. The only people I remember as being generally disliked were some male proctors who served with their wives in large halls as substitutes for parents. Most of us disliked the night watchmen: Snake Pearson, Unk Williams, Max Operman, and a creature named Harris. And I despised old "Peanut Head" Burns who directed the Special Labor Gang to which I was sometimes assigned for using profanity—words that this coal miner's son had picked up along the motor tracks at Spruce Knob.

The most respected and beloved of all the people at Mooseheart were the Mooseheart Mothers. Mrs. York, "Maw" Jones, Mrs. Herbert, Mrs. Van Sickle, Mrs. Davidge, Mrs. McLucas and our mother Mrs. Rhodes. Dorothy's Album includes a picture from the cover of the *Moose Magazine* for November 1938 showing Blanche Rhodes seated in a chair at work on a piece of cloth while a small child recites evening prayers at her knee. The picture presents Blanche as the Iconic Mooseheart Mother--a secular icon: in a circle of warmth and nurture. (Photo 2)

Photo 3-- Blanche and New Kids at Mooseheart, August 1926. Blanche is centered in the back. From left to right—the four smaller boys in the front row are Clyde Alan Rhodes, Angelo Pastor, Paul Evans Potter, and Robert William Merklein. The six larger boys in the second row are from left to right: Ernest Lloyd Rhodes, Everett Arnold, Clifford Lawrence Rhodes, Louis Pastor-- peeking over Clifford's left shoulder, Gregory Silvano Pastor, and Orville Merklein.

Early in August 1926 we were released from Detention at Walnut Cottage into the City of Children. (Photo 3) We became a part of a community of thirteen hundred children. We joined what was perhaps the largest family of well behaved children with the most enlightened parents living between 1913, when Mooseheart was founded, and 1939 when Blanche Rhodes's youngest son, Clyde, graduated.

According to a pamphlet commemorating the Silver Anniversary of the founding of Mooseheart, no graduate of Mooseheart had ever been convicted of a felony and sentenced to a penitentiary. Subsequently, I heard of only two kids who lived at Mooseheart and were convicted of felonies. There was a rumor of one other but I did not know the person. We were good Kids.

Three of the boys admitted July 26, 1926, would become classmates and team mates. Orville Merklein, Louis Pastor and I played on the Third Team, moved up to the Second Team and finally earned a place on the Varsity football team. Louie and I became keen rivals and close friends in the classroom and on the football field and running track. He was both a good student and an outstanding athlete. I was his equal in the classroom. But the best I could do in sports was to block for him on the football field and finish second to him on the running track. Louis once admitted that he wished the Mooseheart coaches had "put a clock on" my brother Clyde to measure how fast he could run. Clyde was considered by some who saw him run in a football uniform to be the fastest runner produced at Mooseheart. I saw him play football only once, and I'm inclined to agree with them. Of course, Clyde was my brother.

But Louie could run 100 yards on a cinder running track in spikes in almost ten seconds flat—that was moving in the 1930's. And I saw him run, chased him in the 440 dash—more than once. So maybe Louie was faster than Clyde. But then, Louie was also my brother.

On the Second Team Louie and I were often paired in scrimmages to block a Varsity player we called "Tiny Tim" because he was so big. He didn't like us and we were not fond of him. Louie and I fine-tuned the technique of the "high/low" to handle the big guy. I would throw myself at the player's legs, hitting him so his knees would buckle. Louie would hit him above the belt. When we timed it right, we put the bigger player flat on his back. It was really satisfying when Coach Gocher smiled at us, and Coach Seeglitz growled at his Varsity: "Run it again." Once or twice during the season Louie and I put Tiny Tim on his back in successive plays. But on a third try, one or the other of us would probably be creamed. It was worth it.

At Annual Homecomings Louis Pastor and I have at times basked in the memory of those scrimmages against the Varsity and our success in handling Tiny Tim. Louie was seriously wounded in France and awarded the Bronze Cross to go with his Purple Heart. I was in the Pacific. As veterans, we grinned with smug superiority when we heard that Tiny Tim spent the war in a stateside office. We heard a rumor that he had been given a Bronze Cross for "the masterful way in which he wrote his reports."

ADAPTING

BLANCHE RHODES' ELDEST son learned painfully but quickly to adapt to Mooseheart. I discovered one must adjust to the rules of two groups with seeming conflicting interests and values. Moreover the rules of both groups were seldom set down in writing, and often conflicted with my ideas of right and wrong. Those differing groups were the Mooseheart Kids and the Mooseheart Officials who made the rules and had the power to enforce them.

My first conflict was with the Mooseheart Kids. One's principles are everything—so I was taught. I always tried to obey my mother—it was a matter of principle, a moral obligation. I discovered, however, one cannot always do what mother tells you to do if you are to survive.

Trouble occurred the first day out of Detention, a miserable day. I was assigned to North Hope Hall, one of the ten larger dormitories for boys, filled with about forty noisy brats ranging in age from six to fifteen. I knew no one. I remember that breakfast consisted of oatmeal, buttered toast, milk and an orange, gobbled up by strange ruffians seated in groups of six or seven at round dining tables.

The meal began with a silent period of thanks for our food—-referred to as "Saying Grace." Awed by the white table cloth and white cloth napkins—the first I had ever seen. [Not long after we arrived at Mooseheart, we shifted to paper napkins.] I bowed my head like the others and silently repeated the words mother had taught us: "Oh Lord we thank thee for this food, our life and health and every good. May manna to our soul be given-- bread of life sent down from Heaven—Amen."

After breakfast I was told to dry the dishes. So far I was in no trouble. I knew what to do. Mother had taught me at Spruce Knob to dry dishes after they were pulled from the steaming water in the rinse pan.

Unfortunately, mother's teaching would land me in trouble. A few minutes after the first whistle for school sounded, I crossed the old steam ditch and started for school. The ditch, about four feet deep and fifteen feet wide, was never filled while I was at Mooseheart. A favorite place for playing, it ran west from the power house past the Vocational Arts building, the one story Laundry building, the two storied Hope Building and the three storied Industrial Halls to Chapter Hall.

Dressed as mother had taught me, I was a real dandy in my dark blue suit. A white handkerchief peeked from the breast pocket, accenting my white shirt. (Photo 1) I wore high topped, carefully laced, Buster Brown shoes and long black cotton stockings, pulled tightly up my skinny legs and over my knees. The stockings were held in place by black elastic garters mother had sewn in my knickers. It was those knickers, pulled above my knees that caused the trouble. Kids who wore their knickers that way were wearing "knee scortchers."

I crossed the steam ditch without messing up my shoes or suit; I pulled my stockings up tightly and secured them with the elastic band in the bottom of the knickers. As I headed toward the grade school classrooms four rows of wooden buildings near the old High School, I heard a mob behind me jeering: "knee scorchers! knee scorchers!"

With no warning a gang of little snots—none larger or older than I was—threw me to the ground, pulled my knickers below my knees--without any explanation, and allowed me to get up.

Dusting myself off and muttering (soundlessly) "dirty little bastards." I straightened my stockings and pulled the knickers above my knees to hold my stockings in place. I looked at the little bastards, turned and walked towards the class rooms.

Again, without warning, they tackled me, threw me to the ground, and pulled the elastic bands in my knickers below my knees. They were not gentle in getting off of me, pushing my face into the ground against the exposed root of a tree. Finally they allowed me to get up and pull the white handkerchief from my suit coat pocket. They backed off and watched as I wiped the blood from my nose and the cut in the corner of my mouth.

"Dirty little bastards," I muttered to myself. I didn't dare call them what I wanted to call them. I knew better. I had learned on the motor tracks to the coal mines at Spruce Knob that a guy was expected "to fight to death" if anyone insulted his mother. Again it was a matter of survival; I decided not to fight to death with a bunch of little bastards by telling them what they really were.

Mooseheart is often described as "the school that trains for life." I never wore my knickers above my knees after that first day of school. I don't remember any of the boys who "piled" me. Later, some of them must have become friends of mine. But none of them ever reminded me of the incident. I once asked Dough Guinn about it. He doesn't remember the incident either, but he agrees that many rules we lived by at Mooseheart were the unwritten rules made by students. They were pervasive, effective, and frequently enforced by "piling" or "stacking."

Many of the customs enforced by peer pressure and supported by student solidarity were beneficial—from my point of view. Every male old enough run or throw a ball was forced by peer pressure to take part in organized sports, especially football and basketball, or blow a horn in the Marching Band. With very few exceptions all of the guys conformed. Those who didn't were ignored. Many, like Earl Guinn, excelled in football and basketball and also tooted a horn in the band.

The pressure and solidarity of peers did not allow an older or stronger guy to mistreat a younger kid. The deterrent was "piling" or "stacking"-- the same medicine I had to swallow for wearing the pants legs of my knickers above my knees. A gang of little guys and middle-sized guys would throw a big guy to the ground and lie on him. It was humiliating and effective. While I don't recall a "piling" or "stacking" incident in which I was involved after that first day of school, I was aware of the practice. It protected me from bullies and prevented me from bullying smaller kids.

Mooseheart students allowed two guys of about the same age and size to settle differences with their fists. The antagonists would find themselves in the middle of a circle of boys shouting, "Fight! Fight! Fight!" The antagonists either backed down or fought it out while the circle grew as the calls "Fight! Fight! Fight!" attracted more kids.

Mercifully, the calls and crowd often attracted a proctor or one of the campus watchmen and the fight ended with warning shouts of "Scram! Scram!" The crowd, including the fighters, wisely scattered in all directions. Fighting was punishable with a trip to the West Farm or Heartview for a week or two of isolation, work and tasteless food.

In addition to settling quarrels and saving fighters from a trip to a Punishment Farm, the student's position on fighting often spared the officials from the problem of administering the punishment.

For me, the students' method for settling quarrels between equals was often a win-win affair. One of my first fights at Mooseheart was with Charles Clowser. I was not, definitely was not, getting the best of the fight. I was saved by the warning: "Scram! Scram!' Charles was satisfied with the fact that he was beating me. I had no desire to continue the quarrel. We escaped punishment, for no official caught us, and most importantly: no one told on us, no one snitched. I remember no instance of a Mooseheart kid ever snitching. We stuck together. It was us against them—authority, officials, petty enforcers--the world.

An incident showing the effectiveness of the code against "snitching" and student solidarity occurred shortly before I left Mooseheart: Tony West and Al Davis broke into the campus grocery store and took something—probably sugar to make candy. The officials were furious and decided to break the no snitching code.

The Supervisor of Boys, "Square Deal" aka "Tack Head" Swasey, put the entire campus on "No Company/No Privileges" status. He dictated that all student privileges be cancelled--there would be no Friday night dances, no Saturday movies, and no trips to Aurora or Chicago until the thieves were either identified or turned themselves in. Every student and every Mooseheart Mother and most of the hired help on campus knew or learned that the "theft" was committed by Tony and Al. But none of the officials with the power to punish were told. No one "snitched."

About six weeks after the sugar disappeared "Tack Head" finally lifted the campus wide punishment. I always suspected that Father (Later Monsignor) Laffey, prevailed upon old "Square Deal" Swasey to withdraw the basically unfair ruling. The good priest understood student values: copping sugar from the campus grocery store or snatching melons from the Mooseheart farm or even gathering hickory nuts from trees on private property was not stealing. Those "reasonable" acts of acquisition were different from taking personal property—nabbing some guy's watch, or sweater, or marbles, or snatching something from his locker. I don't remember losing any of my possessions or stealing anything from another kid. But I did a lot of "copping" and "snatching"--melons, hickory nuts, and a terrier dog that belonged to the people who owned the private property adjoining Mooseheart—the dog got away when I took it with me on another trip onto "private property."

For a short time I with others and made a game of shoplifting at the five and ten cent stores in Aurora. I recall taking a blue neck scarf and other things. And strangely, shortly after one those sprees, I read a magazine article which said that unloved and lonely children often stole bright useless things to feel better. I was ashamed and furious because I knew that I was not unloved or lonely—I was no damned orphan. Perhaps I stopped five and ten cent stores "copping" because I was afraid of looking stupid in the eyes of other kids for being so stupid as to get caught. In any case I reformed.

Peer pressure worked so well and student solidarity was so pervasive at Mooseheart because the majority of us wanted to belong, to fit in, to have friends. For instance I wanted so desperately to be a part of the group that I asked to join the Cadet Corps before I was old enough to wear long pants and be drafted into "drilling." Some students laughed at me; they believed marching around with a make believe gun was "stupid." But they didn't pile me as they did that first day of school for wearing "knee scorchers"— for nonconformity, for defying the group.

I began to sense the unwritten values of the Mooseheart kids and to understand peer pressure and the power of student solidarity. I learned one could usually march to his own drum—providing he moved in the general direction his peers were going. For instance, I regularly wore a skater's woolen cap, similar to those shown in pictures and cartoons of the Devil. I believe I was the first kid at Mooseheart to wear a skater's cap. At that time most of the guys wore plain woolen caps, toques, or went bareheaded.

COPING

IN ORDER TO cope with the rules of Mooseheart officials, students had to know what the rules were and understand how they were enforced. The most important rule was directed at the officials themselves and the hired help. Corporal punishment was not permitted. And with one exception, I know of no child struck in anger by an adult while I was at Mooseheart.

Scotty Cameron, one of my buddies, was caught sneaking out of Middle South after bedtime and struck by a night watchman. The assault was reported immediately, and as I heard the story that watchman never again saw sunrise on the Mooseheart campus. Scotty was not punished.

I witnessed a form of physical sadism that was never reported. "Peanut Head" Burns ran the special work gang for guys whose misconduct was not severe enough for a trip to one of the punishment farms. Peanut had a habit of locking his arm around the neck of some guy who displeased him and grinding his fist into the kid's head. I saw it more than one time. It was said to be painful. Peanut's habit was never reported and he was at Mooseheart when I left years later. We were at fault, I suppose, for sticking so firmly to the code of not snitching that Peanut was not fired.

Mooseheart students were originally controlled by a demerit system. By the time our family arrived it was in trouble. Superintendent Matthew Adams was experiencing serious problems, trying to punish a quarter of the population of the Child City, all those who received more than ten demerits a week. Albert Bushnell Hart, a Professor of History at Harvard and a member of the Board of Mooseheart Governors, heard about the problem. Old Bushy, as the kids called him, laid the law down to Superintendent Adams and the situation began to change,

Less than a year after we arrived, Ernest N. Roselle succeeded Matthew Adams as Superintendent, May 1, 1927. Not long after Mr. Roselle arrived the demerit system was replaced by a merit system. It didn't work. The details were overwhelming, much too complicated for the proctors and matrons. Some were directed to evaluate each week the degree of citizenship of forty teen-aged boys, on a scale of one to ten, according to everything from Attitude to Loyalty, to Reliance and Self-Control, through Trustworthiness to Zeal. It just did not get started.

Yet, the failure of the merit system was probably the best thing that happened at Mooseheart after the cornerstone was laid in 1913. A most important part of the merit system survived: "All students were assumed to be good citizens entitled to all of the privileges of the City of Children."

We were no longer inherently bad children to be caught and labeled with demerits and regularly punished. After the merit system collapsed under pounds of paperwork, episodes of serious misconduct were handled case by case. House parents, teachers and other minor officials reported such misbehavior to the Supervisors of Boys or of Girls. Serious misconduct included running away and extraordinary insubordination.

Routine misconduct was seldom discovered or brought to the attention of the administration, that is--the Supervisors of Boys or the Supervisor of Girls. We were seldom caught and punished for leaving the campus without a permit, using profanity, fighting, and offenses involving smooching, petting, or being alone where boys and girls should not be alone together.

Students involved in extreme cases of running away and blatant insubordination, and those seen as chronic malcontents, were warned by Superintendent Roselle and given an opportunity to "straighten up." Repeat offenders were likely to be given a small trunk for their possessions and a one way ticket back to the Moose Lodge that sponsored them. We said they were "bounced out" or "shipped home." But that was rare. Students whose mother remarried or those who wanted to go live with relatives also left.

A proctor or matron could deal out minor punishment for minor offenses such as being tardy, or leaving the bed sloppy or horsing around during study period. A kid could be denied a ticket to the movie (which was seldom collected), or denied a pass to fish on the far side of the lake. Minor offenses dwindled as off-campus house parents were replaced by Mooseheart Mothers who were respected, obeyed, and loved.

I recall only one Mooseheart Mother who was disliked by most of the campus. Old Hatchet Face took it as her duty to snitch on couples trying to smooch in the darkness as they left the dances or movies. Her name drew derisive laughter when she was mentioned in a recent Homecoming address.

She snitched on Ginny Generis and me for smooching after a dance. We got off easy with only two weeks of No Company—it was our first offense. Couples caught smooching were usually given thirty days of No Company. Their names were posted on the campus bulletin boards in every hall. The rules were strictly enforced: No smooching.

It was seldom mentioned but clearly understood that sexual intercourse would result in the offending couples being "bounced out"—sent home to their sponsors. I know of no instance of sexual intercourse involving students while I was at Mooseheart. Yet it was once rumored among a few older students that one couple had been able to get into the channel under the sidewalk that carried utility pipes. There, according to the rumored, they explored and enjoyed their interesting differences. I considered the rumor a fantasy born of wish.

Student solidarity and values enabled many kids to ignore the rules. For instance Nunzio Ferrara and I were clearly breaking a serious rule every morning for more than three weeks. We crawled out a window at Krebs Hall (while Mrs. Jones slept) and ran around the lake as fast as I could run. Nunzio, who ran the hurdles, was training me to beat Coach Seglitz's quarter milers in the Junior-Senior track meet. No one snitched and no one caught us. No one was hurt!

I was breaking a rule when I crawled out a window shortly before midnight to carry my sick pup to Father

Laffey for help. None of those who knew of the incident snitched —including our good priest. He didn't even chide me. And I didn't belong to his flock.

Ginny Generis and I, with Meg and her boy friend, (later Meg and I, with another couple) were all involved in smooching on the girl's stairs in the high school building while classes were in session. Students saw us. No one snitched, although Ginny and I quit "going together" and Meg and I became a pair. Then Meg and I quit and I started going with Ginny's sister Prudence. While hurt pride and jealousy figured regularly in our complicated romancing, no one snitched. None of us were put on No Company—denied our privileges, forbidden to walk together to classes, or to attend dances or movies because another student snitched.

However, I was caught three or four times and punished for misdemeanors. One type of offence was using profanity and vulgar speech, a habit I had picked up along the motor tracks at the mine. As punishment, I was sent to Peanut Burns' Special Labor Gang. The Gang met every day after vocational training and during free time on weekends. The Gang had to sweep the all-purpose hall in the Roosevelt Memorial Auditorium, to set up up the chairs for Assembly, to rearrange them for Friday night dances, for Saturday movies, for Sunday Catholic and Protestant worship services, and for the Sunday afternoon concerts. In addition, we policed the grounds, picking up litter with a nail driven into the end of a broomstick, then depositing the trash into gunny sacks hung around our necks.

I was lucky enough to escape a trip to the punishment farm by heeding the warning shouts "scram, scram," which ended one of my fights at Mooseheart. But another fight with Jerry Van Sickle ended badly for both of us. He was smaller and faster than I was, but I was holding my own. We were too interested in destroying each other to heed the students warning "scram, scram!" Pegleg Clawson collared us and sent us to Captain Quick. He sentenced us each to a week at Heartview, one of the two punishment farms. There was little difference between Heartview and West Farm except that Mr. Prentice and the formidable Mrs. Prentice were the more unpleasant pair.

The routine was the same for both farms. Each morning, up at 6:30, we ate our "breakfast" of yellow cornmeal mush, seasoned with black molasses and covered with milk. We also had a glass of milk—Mooseheart kids were expected to get on the outside of at least a quart of milk a day—from the Child City's own dairy.

After breakfast and assignments, we marched two abreast back to campus to attend classes until noon. We assembled for lunch in one of the vacant classrooms. We ate a sandwich made of butter between two slices of brown bread and a second sandwich of baked beans between two slices of white bread. These were accompanied by another portion of the quart of milk we were supposed to consume for the day. Charles Shepherdson, class of 1934, says the first sandwich was made with jam. He was at Mooseheart before our family arrived, back in the demerit era. So Shep should know, for he spent most of his time at one of the farms until he reached the eighth grade and the demerit system was scrapped.

We spent the rest of the afternoon in our several vocational training classes and then marched two abreast back to the farm. At the farm we did useless work until supper. It was the one real meal of the day: heavy on bread, meat of some kind, vegetables, milk and perhaps a piece of fruit for dessert. After supper and our kitchen assignments we were seated in straight backed chairs and "allowed" to read our King James Bible or the Catechism for thirty minutes. Following a study period, reading in silence for an hour, we crawled into bed.

Weekends at the farm were made miserable with make-believe work. But the real punishment was missing our friends, as well as all football and basketball games, all sports practice sessions—which could cost a guy his position on the team. We also missed the Friday night dance, the movies, and the usual weekend freedom from classes and vocational training. Sunday we did march to campus for Church services and immediately marched back to the farm when services were over.

After the merit system replaced the demerit system the population at Heartview and West Farm dwindled. There were never more than forty guys at the two punishment farms during any week—that's all they could house. No new farms were added between 1918 and 1939. The two punishment halls, Fez and West Legion were closed about the time the demerit system was scrapped. I was at Heartview once for a week and at West Farm once for another week.

Earl Guinn, my "adopted" brother, my best friend, and later my brother-in-law, believes student solidarity and values coupled with an intense campus-wide interest in sports and music programs account for the decline in student misconduct and the emergence of the unique

Mooseheart spirit between 1919 and 1939, the years while someone in our family was there.

Generally we were "good citizens" and followed the rules. As children do, we sometimes broke rules, when we thought we could get away with it. When we were caught, we were punished. Often we were successful and avoided punishment. No one was hurt. In that environment, our family coped and began to thrive.

THRIVING

LET'S IMAGINE THE sort of things that might have happened on a day during the week of my 14th birthday in 1929. By then Mother had accepted Earl Guinn as her fourth son. Earl, Cliff, Clyde and I were all living in Middle South and thriving. Sister Dorothy had disappeared into a hall on the girls' campus. We saw her often when we visited our mother where she often worked in West Virginia Hall. Dotty appeared happy. With her brood about her, Mother began to cope by the spring of 1927.

Shortly before the six-thirty whistle blew, I went to Clyde's bed and waked him. He wanted to get up early and beat the crowd to one of the three showers. Some of the big guys had started snapping each other (and some of the smaller kids) with the wet ends of their bath towels. The towels snapped like firecrackers and could raise a painful welt on a bare bottom.

Clyde didn't like it and could have gotten his older brother into a fight. Clyde was not above smacking some guy with his little fists—some guy bigger than the pair of us could handle. Discretion, one of the keys to survival, prompted me to get Clyde up and dressed before the big slobs wakened. Neither of us wanted any part of their silly shower-room games of grab-ass and towel snapping.

We made our beds, showered, brushed our teeth, dressed, and combed our hair. We were in the locker room where I was showing Clyde how to throw a football, when the six-thirty whistles sounded to wake the campus.

Clyde was a quick learner. He could lay the football on his little hand, sense where the seams were, and zip the ball the length of the locker room with accuracy. Years later, he would be elected to the Mooseheart Athletic Hall of Fame. He earned his way through Lawrence College, in Appleton, Wisconsin, running, kicking, and passing a football. Times we spent together as kids created life-long ties of affection and respect, strengthening our bond of blood.

At six-forty-five when the first whistle blew for breakfast, we put the football in my locker, in my dirty laundry bag. I had chosen that football for my Christmas present last December. I didn't hide it because I was afraid someone would steal it. Some of the bigger guys would just take it, and tease me with it when I tried to get it back. Eventually they would return it. I just didn't want to play that game.

After hiding the ball we went into the kitchen. It was Dough Guinn's week to make the toast and Cliff was assigned to butter it. Mrs. Bozarth, the mother of Frank Bozarth, was the cook. She didn't pay any attention when Cliff soaked a piece of toast in butter and gave it to Clyde. Clyde broke it two and gave me half, Earl whispered "Get outta here before Mrs. Herbert catches us." Mrs. Herbert, one of the most popular matrons on campus, was a Mooseheart Mother. She was a big strong woman who wouldn't put up with a lot of nonsense. Earl loved her. I liked her, but I was also half afraid of her.

Chewing on our buttered toast, Clyde and I went into the living room to see if we could find a copy of the Chicago Tribune sports page. They sometimes carried a write-up about Mooseheart football. We did find a copy of the Aurora Beacon News sports page and a little squib about the Third Team's Red and Black Football game ending spring practice. None of the players were listed. But I was sure I would make one of the squads and get on the field for a couple of plays in the fourth quarter. Coach Johnny Williams always tried to give everyone a chance to play. However, I began to have an uneasy feeling that after two years on the third team, suspecting that I wasn't cut out to be a football player. Even so, I held onto my dreams.

When the seven o'clock whistle blew for breakfast Clyde and I were behind our chairs waiting. Mrs. Herbert came in and looked around to see if all thirty-seven of her kids were present. Satisfied, she bowed her head as signal for us to bow ours in silent thanks for our food. After what seemed time enough to give thanks for every orange beside every kid's cereal bowl, she raised her head and seated herself. The kids scrambled into their chairs and gobbled up their cereal, doused with milk and sweetened with sugar. Each of us also had two pieces of buttered toast, a large glass of milk and of course our own orange or an apple.

Most of us finished our breakfast and had been excused from the table by the time the whistle blew to end breakfast at six-twenty. Each one carried his dishes from the table, cleaned them and piled them up to be washed. All of us had an assignment to complete before leaving for school. I tried to finish breakfast as soon as possible to finish my job, setting a table for dinner. Then I could dress for school.

At Mooseheart, personal appearance became important when a guy reached fourteen. Everybody said so, the Matrons, our teachers, my mother and the coaches. We were actually supposed to wear a white shirt and necktie on football trips even to dinky places like Oswego. The newspaper and magazine ads always stressed the importance of using Lifebuoy soap and Listerine mouthwash to keep from having body odor and bad breath. While guys I knew didn't seem to pay much attention to their hair, I certainly did. I was one of the first kids to go to Aurora and buy Brilliantine. I shared it with some of the guys who parted their hair in the middle and plastered it down like mine.

By the time the first whistle blew for school, I was out of the door and down the steps. I ran into Charles Shepherdson, one of the kids I lived with when I was in Upper South. Shep had moved to Chapter Hall, which was a privilege hall for older guys. He had been a skinny little kid, not big enough to play third team football. He became our manager and water boy last year for the Third Team. Shep didn't go out for the third team this year—even though he had grown up like a weed. Instead he went out for basketball and made the varsity although he was only a freshman. I don't remember any one else in Mooseheart athletics making a varsity team as a freshman, although John West and Clem Faust were pulled up from the Third Team to play on the varsity football team. Shep's athletic ability was so remarkable that he stepped in and played quarterback on the varsity football team during his last two years at Mooseheart. He was so good he didn't have to pay his dues in the mud on either the Third Team or Mr. Gocher's Second Team. He had one weakness; he could not throw a forward pass where I could catch it.

Shep and I stopped at the Power House to watch the big guys walking to school with their girls.

"You don't have much of a chance with her," Shep said when he saw me looking at Prudence Generis and that creep carrying her books. I tended to agree with him.

"I like her sister Ginny," I told Shep.

He just grinned, like he knew better, as we pushed into the crowd behind Prue and the creep. Shep was younger than I was so we parted to go to our assigned classes when we reached the school building. I was in my seat in the Chemistry/ Physics Lab when Mr. Gocher started talking about Boyle's Law. I spent the second period in the Library working on a paper for Miss Hance's Civic Class. The third period I went to Mr. Wahl's class in Math. He took much of the period trying to explain that the ratio of the circumference of a circle to its diameter is 3.141592+. I left the class at eleven fifty-five for noon recess and dinner, still puzzled about infinity. How could something just go on forever—without coming to an end?

I don't recall what we had for dinner but the food was usually good. I truly enjoyed most of what was put before me. And if I didn't want to eat something, I could always find something I liked to fill me up. So let me describe one of the few meals that was memorable for being taste-less: creamed codfish balls, canned peas, perfection salad (that is, shredded cabbage, carrots, and onions in Lemon Jell-O), and tapioca pudding. It was nothing much. Of course it was a balanced meal accompanied by the usual glass of whole milk. Miss Myrtle D. Barthlomew, a trained dietician, prepared the menu and was much better than the person she succeeded.

Miss B liked me because she was a close friend of our mother. But I could never understand why she often gave us something as distasteful as creamed cod fish balls, just because they were supposed to be "good for you."

Dorothy found a copy of Miss Bartholomew's Mooseheart Cookbook among our mother's possessions. It begins with the assertion:

> "A good cook is accurate, neat, particu-lar, resourceful, and recognizes the value of the appearance of food as well as good preparation."[She elaborates each point and concludes with the admonition:] "The successful cook never puts recipes together by guess. . . ." [Her book con-tains recipes, instruction for the prepara-tion of food of all kinds as well as Safety Instructions which begins with a warn-ing:] "Roaches and flies are not tolerated in any clean kitchen. They are disease carriers." [I do not recall seeing roaches at Mooseheart or many flies in the kitchens there.]

After the noon meal, I finished my assignment cleaning the bathroom. It was a lousy job but I would get another when the new assignments were posted in couple of months. I went directly from the Middle South to Mr. Elwood's Drafting class in the Industrial Arts Building at one o'clock. I finished that course with useful skills and an enjoyment of planning that has kept me owning a drawing board, a T-square, a sharp pencil and a soft rubber eraser throughout my life.

My next class was in painting, beginning at two-thirty. It was the sorriest class I ever took at Mooseheart. We spent our time varnishing scuffed up hardwood chairs. I learned nothing but the fact that I would not take house painting as a trade. Ironically, twenty-five years later I began an eight year stint teaching students to build and paint theater scenery, while I worked on my Ph.D., specializing in The History of the Elizabethan Stage. Throughout my life I have used skills and attitudes drawn from Vocational Classes at Mooseheart.

Vocational Classes ended at four o'clock with a whistle to begin "Free Time." We were free until six o'clock, unless Assembly was being held in the Roosevelt Memorial Auditorium. Assembly brought all the Mooseheart Kids together (except the kids in the Baby Village). I usually enjoyed Assembly because we often had visitors who spoke to us: sports celebrities like Knute Rockne, Hack Wilson, and Bob Zupke, or movie actors like Mae Murray, and Wally Ford, or African Explorers or Missionaries. During Assembly we would also receive information about our everyday activities, and as Mr. Roselle or one of the officials droned along, I could often snatch a glance at Prudence Generis.

Since there was no Assembly during the week of

my 14th birthday in 1929, the period between four and six o'clock was devoted to the usual free time activities, including practice for the several athletic teams. It happened this week to be the last scrimmage before squads were selected for the Third Team's Annual Red and Black Game. Gee! Maybe I would be picked as a substitute for one of the squads. Maybe I could get a chance to show Coach Williams how I could move a tackle in or out—wherever I was supposed to move him. Or maybe I could break through and put the fullback on his tail. Or even better, perhaps I could get down the field, behind the half back and snag a pass—if the quarterback would just put it out there where I could catch it.

We were out of our football gear, out of the cold showers, dressed and back in Middle South in time to sit down for supper at six o'clock. By six forty-five we finished eating, and had completed our assignments-- cleaning up the dishes.

'When the seven o'clock whistle blew we were sitting in stiff-backed wooden chairs around the living room wall. Each of the thirty-seven guys in our Hall "studied" his King James Version of the *Holy Bible*, or his personal *Catechism*. This took fifteen of the longest minutes of my time every Sunday through Thursday night. Yet twenty-five years later, taking classes in English Literature, I would be glad that as a young person I had been exposed the poetry and history of the King James Version of the *Bible*.

Finishing Bible Study at seven-fifteen o'clock I rummaged through books by James Oliver Curwood, Jack London, and settled on *Call of the Wild* which I read and sometimes reread. The period from seven-fifteen to eight o'clock was supposed to be devoted to homework. But as long as I was at Mooseheart, I do not recall an assignment involving homework. The time between eight o'clock and bedtime was always "Free Time." We could read, listen to the radio, play indoor games such as checkers or chess, or practice a musical instrument. Actually we could do anything we wanted to do as long as we didn't create a disturbance. Even so the guys practicing on their horns didn't do it quietly and often drowned out the radio.

We all crawled into bed by nine-thirty when the final whistle sounded. I do not recall that any of us sneaked out of the hall or were involved in a pillow-fight at the end of this particular day during the week of my fourteenth birthday.

My reconstructed account of a typical day at Mooseheart may suggest that we had little or no free time to play outdoors. This was not the case. Our lives were rigidly scheduled and every part of the hour carefully accounted for. But the schedule allowed us long periods of "free time" on the weekends, holidays, and during two weeks of summer vacation in the dog days of August. During Spring and we were free from the time we finished our assignments after supper until fifteen minutes before dark when the first whistle for Bible Study sounded.

Free time at Mooseheart was gobbled up by football, baseball, basketball, indoor and outdoor track meets, swimming in the Mooseheart Lake, boating, fishing, and ice-skating. We could also use free time in musical activities like marching bands, orchestras, and glee clubs. We participated in the Friday night dances, Saturday night movies, Sunday morning worship services, Sunday afternoon concerts and a kind of military drill that would develop into an ROTC program after I left Mooseheart.

During the seven years I was a Mooseheart student, every kid there had a place where she or he was supposed to be, and something he or she had elected to do, every minute of every day—even during "free time."

As soon as Blanche Rhodes' children began to understand what they had to do, what they wanted to do, and what they could get away with, they began to grow and develop—to thrive. We had shelter, plenty to eat, clothes to wear, tasks to perform, training in vocational skills, opportunities to learn, purposeful direction, time to play, room in which to make mistakes and profit from them, friends to cherish. We were considered to be good children, and treated as if we were. And we were never struck or verbally abused by an adult. I believe that few children, even children in wealthy homes, were as fortunate in the 1920s and 1930s as we were.

BONDING

I HAVE ALWAYS NEEDED a close friend--a buddy. My first buddy at Mammoth was my Granny Pritt. At Spruce Knob Emroy Holic was a buddy. Deward Booher was the closest friend I had at Richwood. We exchanged letters for a year or more after I went live at Mooseheart. Wister "Whisky" Reynolds was my buddy while I lived in Upper South. And for a short time I ran around with Scotty Cameron in Middle South. When I moved to North Legion, Lloyd (Bones) Bienvenue became a friend and confidant. I admired him as a good ice skater and a musician—he was an oboe player

None of us are sure whether our family adopted Earl (Dough) Guinn or he adopted us. He was the born in Scammon, Kansas to Alta Alberta Hayes Guinn and Perley Wright Guinn on April 13, 1913. Both of his parents died within days of each other during the influenza epidemic in 1918.

He came to Mooseheart in 1919 with his younger sisters Carolyn and Alberta. He often told us about how Mooseheart took in so many kids during the flu epidemic that he had to live for several weeks in a tent on the lawn outside the Assembly hall. The population of the City of Children peaked to more than thirteen hundred and held near that level until Clyde, the youngest of our family, graduated in 1939.

Blanche Rhodes wasn't happy when she first noticed Dough Guinn. "That Earl Guinn," she complained to me, "is always getting my boys into trouble." Mother had it just exactly wrong. We might sometimes get him into trouble—but that was rare. He was always an easy-going person who carefully followed Mooseheart's rules as well as the unwritten customs of his peers.

Earl says we became friends when he injured himself in a scrimmage between Johnny William's Third Team and Batavia High School's varsity in 1929. Dough says Batavia's big fullback ploughed into him and broke his collarbone. Dough explains that one-armed Hank Jung pulled the jersey over his head rather than cutting it off because the jersey was irreplaceable. After all, I was there to replace Dough on the football team.

As I remember it, the incident occurred in a game with St. Mary's High School in Woodstock. I replaced Earl at Left End for about four plays. It was my first football game. And on the very first play the Woodstock back turned my end and ran by me for a long gain. He was so fast that I didn't have time to blink. Coach Williams didn't have time to jerk me out of the game and send in another Left End. The Woodstock team knew a good thing when they found it: they sent the same little rabbit-back around my end again and he scored a touchdown.

Coach Williams found time enough to send in another of his Left Ends and I found a seat on the far end of the bench. Thus my second season on the Third Team began.

Dough earned his letter, his Peanut, a gold football

with a red 3 in the center, because he was injured and out for the season. I would have to wait for another year to get enough playing time (one quarter in each game) to get my Peanut. After three years on the Third Team, I finally earned my letter—it was promptly stitched onto Virginia (Ginny) Generis's white sweater. I had a girl and she had a football player. Ginny and I had arrived—socially.

Earl was two years older than I was and moved up to play on the Second Team for Coach Gocher in 1930. Dough was an accomplished athlete. I was by that time beginning to be bookish—to compensate for my lack of athletic talent.

Football was at the center of life for most Mooseheart guys, including me. The way to be "somebody" at Mooseheart was to play football or earn a place in the Marching Band. Earl carried a Tuba in the Marching Band. He played baseball, ran on the track team, earned letters on the third team, the second team, and the varsity football teams. He was Captain of Johnny Williams' junior varsity basketball team and was for a year on Carlos Powelson's varsity team. He was elected to Mooseheart's Athletic Hall of Fame and has a plaque honoring him— on the same wall in the Field House as the plaque for our younger brother Clyde Rhodes.

Earl was usually shy—sometimes I've accused him of being prim. I tended to be a profane, irreverent, loudmouth. I have often wondered how two boys so obviously different could become friends—brothers. Perhaps we each needed a brother. Earl had none and Cliff and I had never bonded. Clyde was too young to be a buddy—as he later became.

In any case Earl and I became brothers even though

he was two years older and moved to a different Hall soon after we met. We would never again live in the same Hall or be teammates after our year on the Third Team. We spent some time together in Machine Shop but were never really classmates. Yet by the time Earl graduated in 1932, Cliff and I, and to an extent Clyde, had bonded with him. For Blanche Rhodes he became a fourth son, one who supported her emotionally and financially until she died in 1966. (Photo 4)

Dough returned to Mooseheart for graduation in June 1933. It seemed strange that he spent more time with Blanche and our sister Dorothy than he did with his own sisters or with Cliff, or Clyde--or me. But then I supposed that a boy's mother is more important to him than his brothers and sisters.

I left Mooseheart in August 1933 to attend the University of West Virginia at Morgantown. I sometimes wrote letters but I did not see or talk to any member of my family until I met with Mother in Richwood in August 1937, a couple of months after Earl and Dorothy married. So I don't know much about their courtship.

From what I can reconstruct, Earl mentioned Dorothy—sent her greetings--in a letter to his sister Alberta. At that time Dorothy and Alberta had become best friends. Then Dorothy and Earl began writing to each other—much, apparently, to Alberta's annoyance. She never really bonded with our Mooseheart Family,

While I can't imagine Dough Guinn writing a love letter, I can't imagine my sister writing a letter that wasn't a love letter. Nevertheless, by the time Earl returned for the Graduation ceremonies at Mooseheart in 1933 they were apparently sweethearts.

Photo 4 – Blanche and Her Fourth Son

Dough enrolled in the Southern School of Optometry in Memphis, Tennessee, in 1934. Dorothy said he proposed marriage to her on the Sun Porch of the New York building where mother was then staying. They apparently agreed to marry when he was established in Optometry.

Dorothy and mother left Mooseheart in 1935 to be with Clifford in Los Angeles. Dorothy enrolled in and finished a course in Cosmetology. Dr. Earl Wright Guinn got a job as an Optometrist in Seattle, and they married on Dorothy's 19th birthday June 4, 1937.

Blanche Rhodes had acquired legally her fourth son, and Clifford, Clyde, and I had another brother. Dorothy had a husband. And the Rhodes/Guinn family acquired a new Matriarch, Carolyn Guinn Hanke, when Clyde graduated and Blanche moved to Cedar Grove, WV to take care of Marcena Pritt in 1939. Carolyn's marriage to her high school sweetheart Robert Hanke and their "adoption" of Clyde fused a family group that I call the Rhodes/Guinn/Hanke Tribe from Mooseheart.

Carolyn was ideally suited for the responsibility. How well I remember her warmth and care while she fulfilled her role as Matriarch for four decades. She was exceptionally diligent in safeguarding anything memorable for her "family," from a baby's first shoes, a third-place ribbon for horse shoe pitching in ninth grade physical education class, or a report card with no grade below C. For instance, when I wrote a scholarly book, I mailed a copy to the Mooseheart Library. With obvious vanity, I poked around the shelves in the library during one of the Homecoming weekends to see if I could find it. It wasn't there. I mentioned the Library's loss of the book to Earl. He told me that his sister, our Matriarch, removed it to a bookshelf in the home of the Superintendent of Mooseheart —to be safeguarded. Carolyn continued as the Matriarch of the family until she died shortly after her husband in 1980.

The current Matriarch of the Tribe is, of course, my sister, Dorothy Rhodes Guinn, whose photographs have been safeguarded over the decades; they now provide the pictures for "A Coal Miner's Family at Mooseheart."

PLAYING

A COUPLE OF DAYS after I moved into Upper South in fall 1926, a big kid snatched my cap as I left the hall. He held the cap above his head. When I jumped up to get it back, he tossed it to another guy. I ran to the other guy to get it. He tossed it to a third person. Finally, I gave up and went over and sat down on the steps in entrance to the building. They tossed the cap around from one guy to another, running up to where I sat to tease me into trying to get it.

I thought, "To Hell with it, they can keep the old thing," I just sat there, wanting to cry from frustration-- but knowing better than to show tears.

Finally one of the guys ran over and stood in front of me and waved it.

I just looked at him.

He sneered, "Sissy," and threw the cap at me.

I sat there making no move to pick it up. I knew they would grab it if I tried to retrieve it.

They laughed at me and trotted off to busy themselves some other way.

I had just been introduced to one of the most common games at Mooseheart: "Keep it Away." I suspect it is a universal game that boys use to test and tease smaller boys and especially new kids.

"Keep it Away," was a kind of initiation. It helped to establish a pecking order. Feisty little kids like Clyde "Babe" Rhodes, my youngest brother, would often hurl themselves at their bigger tormenters to retrieve their property. Fighting back established them as someone to respect.

"Keep it Away," sometimes paired up a couple of equals in a fistfight. Sometimes it created alliances when a bigger boy would come to the aid a smaller boy. And occasionally it could set a pack of smaller kids into "stacking" a bigger kid—tackling him, throwing him to the ground, piling on him, and holding him down.

In my case, opting out established me as a "sissy," or at least a guy who wouldn't fight a bigger guy. And they were right. I was not inclined to get into a fight with someone—unless I thought I could "take him"—beat him up.

Mooseheart kids often improvised games similar to "keep it away"--games in which anyone could play, on either side, as long as they wanted to play.

"Throw it Up" was one of the more popular games. Someone would toss a football into a crowd. The guy who caught it would tuck it under his arm and run. The others would chase him, tackle him and throw him to the ground. He would get up after the crowd downed him. Then he threw the ball up or passed it to another guy who caught it, tucked it under his arm and ran until he was cornered and downed. Some of Mooseheart's football stars began by playing "Throw it Up."

For a time, a variation on hockey called shinny was a

popular game. It was played with clubs or sticks used to knock or push a ball or a block of wood around.

The goals were often the roads at either end of one of the large open fields at Mooseheart. Anyone could play, on either side, or change sides, as long as they had a club and wanted to play. The only rule was: "shinny on the right side,"-- approach the ball or puck as a right-handed batter would approach it in baseball--or run the risk getting your shins cracked.

When the Mooseheart Lake froze over we played a variety of ice hockey. We shoveled the snow aside to create a rink. Some of us purchased shoe skates and others even had a hockey stick. Sometimes we used a puck, but mainly it was a wooden block. Occasionally we would choose up sides to form teams. The Mooseheart Lake provided year-around recreation, swimming when it was warm enough as well as fishing. (Photo 5)

Occasionally we would play softball or baseball, using one batter and nine or less defensive players. When a batter struck out or was thrown out the catcher became the batter and the pitcher became the catcher. Defensive players then rotated through the line-up, as the first baseman became the pitcher.

Strangely, we did not often play soccer. I understand, however, that it became a team sport at Mooseheart after our family left.

We often played a form of sidewalk tennis or sidewalk ping-pong by bouncing a rubber ball on a court defined by blocks on the sidewalk.

I have seen bigger guys pitching pennies at a crack in the sidewalk. The object of the game was to get one's penny closer to the crack than any of the other pennies

Photo 5: Swimming in Mooseheart Lake

and claim all that were thrown. This particular game was frowned upon by Mooseheart officials as gambling. Those caught or reported for "pitching pennies" could be sent to the Supervisor of Boys and given a tour of duty on "Peanut" Burn's special work gang.

Jacks would appear on the living room table or the floor in cold or rainy weather.

Many a "cherry" (a small red glass marble) changed hands at one end of the locker room as kids rolled "emmies,' cheap clay marbles from the other end of the room in a difficult gamble to strike and claim the little red marble.

Among the indoor pastimes were contests between guys to construct the strongest spool tractor. These toys were made from used sewing machine spools. A rubber band was fastened with a tack at one end of the hole in the spool. The band was then pushed through the hole and attached to a small stick. A wooden matchstick was often used. Or a stick from a candy sucker could be used. The stick was twisted to create tension in the rubber band. A bearing, whittled from a cake of ivory soap, was often placed between the stick and the surface at the end of the spool to keep the band from unwinding rapidly. Two tractors were placed against each other and released. The tractor producing the stronger force from the twisted rubber band and the stick pushing against the floor would move the other tractor backwards and sometimes overturn it. Often the tractors were placed on the floor to see which tractor could travel faster or farther. The raised edges at each end of the spool were often notched to give the toy more traction. We were always searching for large sewing machine spools and suitable rubber bands.

In addition to improvised games and pastime activities, some kids found the money to support hobbies: building crystal sets (radio receivers) or raising rabbits or pigeons.

It was rumored that a crystal set could be made with a piece of copper wire for a tickler and a cut potato for a crystal. I never succeeded in getting it to work.

However, with a pair of Murdock headphones for a starter, I did build and upgrade half a dozen crystal sets, experimenting with coils and variable condensers. I used my bedsprings for an antenna—and a steam pipe to the radiator for a ground.

Ready made sets and parts for crystal sets could be bought in the five and ten cent stores in Aurora. Most any Mooseheart kid could afford to build a crystal set—or buy one. I could. I had a mother who worked at Mooseheart and made twenty dollars a month.

I should interrupt here to say that after decades have passed, I realize how utterly selfish and thoughtless I was to sponge money from her. Never once did she protest about my requests for money—yet I now believe I left Mooseheart with more money in my student bank account than she ever had in her purse or her bank account--if indeed she had a bank account.

With the crystal sets and parts that I bought with money sponged from my mother, I could pick up WJJD with its transmitter in the Woman's building on the Mooseheart Campus. I could also pick up WGN in Chicago as well as WLW in Cincinnati and KDKA in Pittsburgh.

My memory may be surging and fading like a signal on a crystal set, but I believe I picked up the "Good Music Station in New York" and a powerful station in Del Rio, Texas that advertised patent medicines. Possibly I am confusing my crystal set's performance with the conventional radio receivers I listened to at Mooseheart.

Mother also supported my hobby of taking and developing photographs. While I was in Middle South I used the linen closet in the hall opposite the dining room as a dark room to develop and print pictures.

I don't recall if Terry Kirk was in Middle South and used the same closet I used for my photography. However Terry learned the hobby at Mooseheart, and was a better photographer than I was—by far. During World War II, while Corporal Terence Sumner Kirk of the United States Marine Corps was a prisoner of the Japanese he constructed a pinhole camera from a cardboard box and made photographs documenting the inhuman treatment of his captors. He guarded his pictures and carried them home when the Japanese surrendered. In his book The Secret Camera, Terry tells about his capture imprisonment, mistreatment and his struggle to survive for 45 months. (La Boheme Publishing, Cotati, CA 1983).

One of the very few big guys who did not play football or play in the band started raising shorthaired blue-gray rabbits. He kept them in one of the shacks on the edge of the lake between the Boy Scout cabin and the cemetery. Other kids also started raising rabbits in the same area.

Around the rabbits swirls a bizarre incident. One morning dozens of rabbits were found dead—their necks apparently broken. The slaughter was reported and the Supervisor of Boys gave us permission to join the campus watchmen in guarding the surviving rabbits. Earl and I stayed out all night patrolling the area. We did not catch the rabbit murderer.

Earl believed the rabbits were killed by one of the campus watchmen. I wondered if a watchman, ordered by a Mooseheart official, did the killing to clear out the unsightly and potentially troublesome area known as the shacks. The problem with such a theory is that no Mooseheart official with power enough to order the slaughter of the pets could possibly have been so twisted and cruel. So the incident remains unexplained.

Earl and I owned one of the shacks and used it to keep warm while we changed from our regular shoes to ice skates when the lake froze over. Once we used our makeshift stove to fry potatoes and onions and make coffee. We created no culinary delights.

At the time of the rabbit murders we kept some pet gophers in our shack. They were not harmed, but we decided to move them to a safer place. Mother gave me

12 dollars to hire one of the grounds workers to put skids under the shack and haul it across the campus and the steam ditch. We settled it on the other side of the steam ditch so we could watch it from Middle South where we lived.

The gophers died. We purchased two pair of racing pigeons from Shorty Anderson who ran a lawn mower on campus. Thus began the R&G Racing Pigeon Loft. We bought other birds and traded some of the birds we raised to other breeders in Batavia.

We shifted the males and females about to get colors and presumably better racing birds. It was simple to break up a bonded pair by separating them and providing each with a new partner. The new pair was put together under an orange crate and within a day they mated and were at the business of raising squabs. The incubation period for the eggs was about 17 or 18 days. Approximately ten days after they hatched the squabs were able to take care of themselves.

Earl and I bought a red checked male pigeon that was said to have been a Heitzman- Logan from a racing loft in Kentucky. We provided him with new mate almost every month.

Another prize bird had flown from Las Vegas, NM to someplace in Northern Illinois. We called him our "Thousand-Miler." The original owner gave us two birds for one of the Thousand- Miler's offspring because it had a distinguishing wattle that he was seeking for his breeding program.

We released our birds after they became acclimated and it was a joy, a real triumph, when they all returned— sometimes bringing with them common pigeons. We called them commies and became skilled at distinguishing the worthless commies from our pure bred homers. We purchased fifty numbered bands with our R&G mark and slipped them over the legs of the pure bred squabs we raised. We did not enter our birds in the local club's races because it was too expensive, requiring a special timing clock, club membership and fees beyond our means.

After Earl graduated in June 1932, I crated our homers up and shipped them by Railway Express to Granny Pritt in Mammoth, West Virginia. She began to take care of them, but her husband Charlie Pritt turned them loose before they became acclimated. Presumably some of them returned to Mooseheart. But I never saw any of them again--unless it was the red-checked Heitzman-Logan. I believe he was perched on the coal tipple at Mammoth when I visited Granny while I was in college at Morgantown. It was clearly a homer—not a common pigeon.

Clyde clings to Cliff's leg in (Photo 6) showing how kids at Mooseheart improvised play. The Industry Building containing six boys' halls is in the background. Middle South is on the left side of the middle level. Upper South is on the level immediately above Middle South. Cliff and I spent about half of our life at Mooseheart in Upper South and Middle South. Earl Guinn, Clyde, Cliff and I were together in Middle South with Maw Herbert for over two years, as I recall.

The amazing thing about playing at Mooseheart was the absence of petty rules interfering with our activities. The place belonged to us to use. We built shacks for our pets where we wanted to build them. No one tried to stop us from drowning gophers out of their holes and penning them up in orange crates—except the gophers: they seemed to prefer dying to living with us. I carried dogs into the hall and washed them under the shower in our bathroom. We dug a high-jumping pit in the middle of the field outside the Hope halls. I never tested the officials with a cat or a snake, but I doubt if anyone would have tried to stop me from keeping such a pet as long as it wasn't rabid or poisonous.

Photo 6 – Clifford, King of Snow Hill

COMPETING

MOOSEHEART'S PARTICIPATION IN interscholastic programs was fiercely competitive, a tradition that began with Coach Ben Oswalt's football teams in 1914. When Oswalt died in 1926, Coach William Seeglitz held onto the tradition until he resigned to take another job in 1935. He had an intense desire to win, but left because Mooseheart did not have an enrollment from which he could draw players big enough to compete in the game as it developed.

Johnny Williams, Seeglitz's assistant, however, carried Oswalt's tradition forward and fine-tuned it in a forty-year coaching career at Mooseheart, lasting until he retired in 1962. Johnny was the diminutive 128 pound quarterback of Oswalt's 1914 team and captain of the 1916 team. Johnny's size, the fact that Mooseheart players often "come in small packages," and the schools limited enrollment, figure importantly in the competitive tradition-- the Mooseheart spirit.

Johnny Williams graduated from Mooseheart in 1920 and was appointed in 1923 to coach the Third Team consisting mainly of eighth and ninth graders. When Johnny took over the Varsity in 1936 his quarterback was a 138 pound kid named James Reed. On the same squad was a 137 pound guard who had played on the bantam-weights. Johnny noticed that the guard was consistently out-running his backs in the 40 yard wind sprints. That speedster, our brother, Clyde, became quarterback of the varsity after James Reed graduated.

What did Clyde Rhodes have that attracted Coach Williams attention? In addition to speed he could pass and kick, skills picked up from years of throwing and kicking a football. And Clyde excelled in another Mooseheart program. He was Major Clyde Rhodes senior officer and the leader of the Cadet Corp. Coaching from the sidelines was prohibited when Clyde played football. All plays and all decisions were usually made on the field by the quarterback. Only the captain of the team could—at his peril—overrule the quarterback. Clyde's ability as a leader, a field general, was noted after he left Mooseheart to finance his college education on a football scholarship:

In SPORTSLIGHT, Tuesday, October 24, 1940, a sportswriter comments upon Clyde's decision-making in leading Lawrence College to victory;

> EVERYONE has his idea of why Lawrence football team did an about face last Saturday. Here's ours. It wasn't Dusty Rhodes actual playing but his willingness to call ONE play which provided the victory. This particular play has been our most consistent, but heretofore this play has been saved for the "touchdown spot." Saturday this play was called as many as six consecutive times without variation

other than right or left formation. The description of this play resembles a repetition of old-fashioned power football. The fullback takes the ball from center and ploughs straight ahead behind the blocker who blasts his way between guard and tackle. Monmouth's inability to stop this play and the necessity of distorting their defense to protect themselves made every other play Lawrence used look like pictures in the coach's notebook.

In summary, Clyde fitted into Johnny Williams' philosophy of Mooseheart football. (Photo 7) Clyde was a leader with speed and skills sharpened by years of interest, practice, and experience in football. I was told by former players that long after Clyde graduated Johnny Williams would turn to his team and ask: "Now, what would Babe Rhodes do in this situation?"

For those interested especially in football, an article in Sports Illustrated discusses in detail how Mooseheart teams have overcome their disadvantages of limited enrollment and small players to beat larger schools from coast to coast. Written by James Poling, the article appeared in November 33, 1954, on p. 22, with title "The Mighty Orphans: the children's home run by the Loyal Order of Moose fields a pint-sized football team that has licked big foes for years." Poling lists some of the teams played, from coast to coast, that were defeated by speed, intensive drill, and the innovative techniques devised by Coach Johnny Williams. Poling reported that Williams' teams could run 150 different plays from nine different

Photo 7 – Clyde – the Triple-Threat

formations—without the use of a huddle. It also comments upon the unusual speed and skill of Mooseheart linemen.

By the time our family arrived at Mooseheart the competitive spirit of the football teams had spread from the athletic program and permeated every aspect of the school's interscholastic programs shaping what I call "The Mooseheart Spirit:"

The Mooseheart Spirit is a collective attitude of the entire school, the student body, the teachers and the administrators. We will win if we use our heads, take advantage of our situation, plan and work together, practice together, and conduct ourselves properly in winning and losing.

It was us against them, all forty eight of the then United States, in interscholastic sports and organized music programs. Mooseheart students were welded together as a single unit made up of smaller carefully organized, well coached and intensively trained athletic teams, concert bands, orchestras, choirs and glee clubs.

Morale, discipline, physical conditioning, time for practice and training were not problems. We lived together, ate balanced meals planned by dieticians, attended school and vocational training classes together forty eight weeks of the year with but two weeks summer vacation and two more weeks to observe traditional holidays.

We were coached and trained by enlightened teachers. Most of them believed that scholastic activities were conducted for the benefit of the student. It was understood that every child that practiced faithfully and tried to contribute to the success of the team or band should be considered a member of the group. Size and talent had nothing to do with it. Mr. Warwick encouraged me to play a tuba for weeks and weeks. I alone decided that I couldn't read or produce a note of music, so I quit practicing and turned the instrument in to try to play football.

We had well organized fan clubs with Loyal Order of Moose Lodges in every large and medium sized city in the United States. Their desire to see their teams play and hear their musical performers carried us to interscholastic events from Malden, MA to Tacoma, WA. Dedicated to the Mooseheart spirit that prized sportsmanship and fierce competition, we always tried always to win and to conduct ourselves to honor our fathers and mothers, members of the Loyal Order of Moose.

The summary below will indicate that our varsity football teams won more games than they lost. But more importantly, they were never disgraced in victory or defeat.

Games Played..386
Games Won 237
Games Lost...129
Games Tied..20
Points Scored by Mooseheart................................. 7116
Points Scored by Opponents...................................3672
Average points scored per year by Mooseheart.......... 148
Average points scored per year by Opponents.............76
Average points scored per game by Mooseheart..........18
Average points scored per game by Opponents 9
Number of undefeated seasons by Mooseheart 11
Number of unbeaten, untied seasons by Mooseheart....9

WORKING

JAMES J. DAVIS, Mooseheart's founder, insisted upon the primacy of manual work over "intellectual work." He writes in his autobiography The Iron Puddler that he went to work in the Pennsylvania steel mills at the age of eleven.

Neither Mr. Davis, nor Homer Rhodes, who went to work in the coal mines when he was twelve, ever required one hour of work from me "for my keep." However both men regularly made it possible for me to work for wages at responsible jobs. When I started college in September 1933, I had saved enough to pay most of my tuition and fees ($37.50 per semester) until I completd my BA degree at West Virginia University in June of 1936. I finished the four year course in two years and nine months by attending summer school sessions, and through an Honor Point system for good grades then in effect at WVU.

As it turned out, none of the funds I took to college came from the pigeon project with Earl. We started it with money "borrowed" from mother to move the loft to a safe location. For a time Earl contributed to the project with money earned operating the campus switchboard. We did not keep careful records. So after he graduated, I liquidated the racing loft with a profit—for myself.

I began a new business by shipping the racing birds to Granny Pritt in Mammoth, West Virginia. Then I turned the loft into a place for "breeding squabs." Actually I assembled a number of different kinds of pigeons: fantails, rollers, tumblers, white mondains, silver kings and a lot of "commies." The business ended before it could really get started.

Miss Bartholomew, Mooseheart's dietitian, asked if I could provide fifty "squabs" at twenty-five cents apiece. She needed them for a banquet in the Campanile for a group of dentists.

I guess those dentists had good teeth because I used most of my breeders and commies to fill the order for fifty "tender" squabs I was supposed to provide. No one complained.

Thus ended the R&G Pigeon Breeders partnership with an apparent profit of 50 cents. Mother had "loaned" $12 to have the "loft" moved from the Lakeside and Mrs. Bartholomew paid $12.50 for my breeding stock— converted to "tender" squabs.

My experience offers a textbook example of how to succeed in business. Mother's loan was not repaid, and I forgot to split the 50 cents profit with Earl. Perhaps I should have left Mooseheart and started a business or set up a company of some kind. Consider for instance: my skill at raising venture capital, my example of how to treat partners, my concern for "quality control," my ability to reorganize an unprofitable partnership, my skill at creative accounting.

Instead of exercising my latent skills in business and finance, I spent much of my free time at Mooseheart working for 15 cents an hour. For a while I dusted every chair, in every classroom in the high school building every Saturday morning. Later I worked at the grocery store filling orders for the boys and the girl's halls. Still later, after I had moved to Quaker City, one of the privileged dorms for boys, I remember cutting weeds on a fencerow north of Mooseheart Lake. The occasion was memorable for two reasons. For the first time, I was earning 25 cents an hour. Secondly, I found a treasure. One of the campus workmen happened to remark that "Indians used to live around here."

"Yeah?"

"Yeah! Everybody knows that," another workman responded. The conversation ended there.

Later during the afternoon, I uncovered a perfect arrowhead. I gave it to mother and she kept it for years. I sometimes wonder if the artifact was planted by one of the workmen to tease me. But when I showed them the arrowhead, they did nothing to support such a suspicion.

Many guys like me worked at Mooseheart doing scores of odd-jobs. I deposited the money or, perhaps, it was deposited for me in the Students Bank. When I first arrived, the Richwood Moose Lodge sent me a few dollars, to begin my bank account. For a time I would draw out a nickel on Saturday morning, buy a bag of sugar at the grocery store and take it to the hall where mother was working as cook or a matron. She would make seafoam candy for all of her kids who showed up for family time together.

One of the ways of earning money was to sit on a bench at the entrance to the Campanile. Mrs. York, the receptionist, would summon one of us to guide visitors.

Often members from Moose Lodges throughout the United States, Canada, and even England wanted a guide to show them the campus. I developed a route and a line of patter to give visitors a thorough picture of the campus and create a good impression about their guide. (Photo 8) The visitors often tipped well-—especially a small group of members of the Moose or a man and a woman with a couple of kids. Normally I would get a dollar or two, or even three, but sometimes only a couple of coins, and on one occasion five dollars.

For members of the Moose, I always began the tour at the statute of James J. Davis on the rotunda of the Campanile (Photo 9). There the Welsh immigrant and iron puddler, Director General of the Loyal Order of Moose, founder of Mooseheart and Moosehaven, U.S. Secretary of Labor under three presidents, and former senator from Pennsylvania, is memorialized by a life-sized bronze statue. We all look reverently at the statue of the man resting his hands on the shoulders of two kids, a boy with a hammer and tools and a book, and a girl carrying a book and a ruler. I would tell them, as if they couldn't read: this is James J. Davis the founder of Mooseheart. He says "No man stands as straight and tall as when he stoops to lift up a child."

About a year before I left Mooseheart, a man from Morgantown, West Virginia, Christopher Droop, and another man came to see us. Mr. Droop said he worked at the Cherry River Boom and Lumber Company store in Richwood and had known my father. I think he said he had been a member of the Richwood lodge, but I'm not sure.

Photo 8 – Aerial View of Mooseheart Campus, ca. 1933..
Unfortunately the athletic complex at the Eastern edge of the
campus is not shown. Also, Mooseheart Lake, many girls' halls,
several cottages, and the Baby Village on the Western edge are not
shown. Most buildings on my tour lie on a diagonal line beginning
at the campanile in lower right hand corner and continuing to the
dairy barn (not visible) in the extreme upper left hand corner.

Mr. Droop seemed like a nice man, about our mother's age.

Mother, Cliff and I had our pictures made on the rotunda of the Campanile with Mr. Droop and his friend. Mother suggested that I show them around the campus. I jumped for the idea—it was a real chance to show off our home and get a good tip. I had the afternoon free; it was a Saturday. So I made up my mind to give Mr. Droop and his friend a real tour with my best line of patter. They could be paper-money tourists.

They didn't want to see the football field, and the classrooms were all closed on Saturday. So, we walked over to the Assembly Hall, which contained the Roosevelt Memorial Auditorium (Photo 10) and Superintendent Roselle's office.

Photo –9 Statue of James J. Davis, Founder of Mooseheart

I took them into the auditorium which we called the Assembly Hall. "This is the most important room in the most important building on campus," I explained. "All of the kids, except the little guys, come here once or twice a week to hear announcements, and to listen to famous people who come to see us. People like Babe Ruth, Knute Rockne, Bob Zupke, and movie stars like Mae Murray and Wally Ford. A lot of famous people--teachers, and missionaries come here to talk to us. I don't remember all their names. But Sergeant York, the World War hero, was one of them."

I explained how the folding seats in the Auditorium were pushed against the wall on Friday nights so the guys in high school or who wear long pants can come here and dance with their girls.

"Do you have a girl?" Mr. Droop asked.

"Naw, we quit. Besides, I'm too busy with football. Sometimes I come to hear our dance band, the Davidsonians, or to see my mom and my sister and maybe dance with them." I didn't want to get started talking to those guys about girls or getting into trouble for "smooching."

I just wanted to get on with my patter about how the auditorium was used. I told them that "On Sunday mornings the seats in here are turned around and church services are held for all the Catholic kids." I thought they might be interested in knowing that "the Catholic kids can brush their teeth before Mass but they can't eat or even have a swallow of water until after Church. Then, when they get back to the hall we can all have breakfast." I could see that I wasn't saying much to interest Mr. Droop and his friend.

Maybe they weren't interested in Catholics. So I moved on to talk about how we had our church services, all of us Protestants. "We have church right here in this place, after Catholic kids have their services. That is, we

Photo – 10 Assembly Hall in Roosevelt Memorial Auditorium

have our services after Peanut Burns and his Special La-bor Gang clears away all the Catholic saints and candles and things. One time Billy Sunday, who used to play for the Chicago White Sox, was here to preach to us."

"No kidding." Mr. Droop's friend said—I sort of thought he was making fun of Billy Sunday, or me or something. He sounded snotty.

Maybe they would be interested in music. So I told them that on Sunday afternoons we usually had a con-cert by our philharmonic orchestra, or one of the Glee clubs, or our Marching band. "Mr. Davis is strong on music and we have bands and singing and all that stuff. Sometimes guys like Art Wright or Ralph Hill sing or Flossie Ford plays the organ—-I pointed to the pipes in the rear of the auditorium.

"Do you play a musical instrument?" Mr. Droop asked.

"Naw, but my friend Dough Guinn does. He plays a

Tuba and he plays football and basketball, and baseball—and he's on the track team, too.

"That's nice," Mr. Droop said again.

I took them outside, led them down a slight incline on the side of the Assembly building, and stopped at the shoe repair shop. "This is where I trade my shoes in when they get holes in them. If Mr. Joe the shoemaker has a pair that he repaired for me, he checks me to see if they still fit. If they don't, I get a new pair. A couple of guys I know are taking shoe repairing for their trade—they always seem to have new shoes," I reported.

"Can you show us the clothing store?" the guy with Mr. Droop asked. "I work in the clothing store in Summersville. You know where Summersville is?" he asked.

"Yeah," I said, "it's the county seat of Nicholas County in West Virginia. But I never been there. My Aunt Belle's boy friend told me he saw them hang a guy over there because he killed someone."

"I saw them hang him" the guy from Summersville said. "They had him all dressed up in a new dark blue serge suit and a white shirt—-but no tie. When they asked him if he wanted to say anything he started saying the Lord's Prayer. After he finished they put a pillow slip over his head, tied a rope around his neck, sprung the trap--and that was that. Sure didn't get much wear out of that new suit."

"Well," I thought, at least Mr. Droop's friend is interested in clothes. So I skipped the other things in the Assembly building: the student bank, the WJJD broadcasting studio, the grocery store, the girls clothing store and took them directly to the boy's clothing store.

Mr. Smathers, who ran the store and the guy from

Summersville, really took to each other. Mr. Smathers explained how each boy's clothing was distributed:

> Take Ernest, here, for example: he has an allowance for say one suit a year. When he thinks he needs a new one, he goes to Miss Pratt in Quaker City hall: She fills out a requisition for a new suit. He brings it in, and after I check his record, we go over the racks together. He picks one out and I issue it to him.
>
> He got a new suit last month, and after he wore it over the weekend he brought it back. Didn't you?

"It was too tight under the arms," I said.

"Maybe his girl didn't like the color? That's more like it, ain't it, Mister?"

"It wasn't like that; it was too tight under the arms," I said, coloring up. I didn't want him to go talking about my girl. Beside, it was my mother who said I ought to get a double-breasted suit, one in a darker shade of blue. She was the one who thought it was too tight under the arms. And I didn't want anyone to know that my mother had anything to do about the picking out my clothes.

"It was too tight under the arms," I repeated.

"Well, you got another one, didn't you, Mister?" Turning to the two men, he smiled and jerked his head and a thumb in my direction. "Customer's always right, around here, ain't that so, Mister?"

"Yeah," I agreed.

"You tend the store," he said, "while I show these

gentlemen the stock we got on hand. Just sing out if any one comes in." They disappeared into the back room.

After a while they came back, laughing and joking. Then they thanked Mr. Smathers and he thanked them.

I took the two men outside the boy's store, pointed to the girls store and explained, "I never been in there, don't know much about it. But my sister, Dorothy, says every girl gets two new dresses each year. They all trade around and wear each other's dresses. And they can buy clothes in Aurora—if they can get enough money. Dorothy says they're always letting out hems and, making over old dresses. She says no girl likes wearing the same dress every day. I'll take you in there, if you want to see it," I offered.

They said they didn't really need to see it. Then I pointed to the students' bank and volunteered: "Most of the kids got their own account and can put money in and take it out when they got it."

"Where do you get money to put in a bank account?" Mr. Droop asked.

"I've helped the janitor in the high school building, worked in the grocery store, worked on the grounds cutting weeds--anything to make a dollar," I said. "And sometimes the Richwood lodge sends our family money."

I headed them back past the grocery store and we peeked in the door. Then we stopped at the powerhouse and looked at the 1923 fire truck. Outside, I pointed to the powerhouse whistle. "It's run by steam. It gets us up every morning, tells us when to eat and go to school and go to bed. It'll go off after a while to tell us to get ready for supper."

We returned to their car near the Campanile. We drove back, passed the administration building, and passed the powerhouse and the water tower.

We stopped and looked through a window on the first floor of the three-storied Vocational Arts Building. "I took Drafting in that room with Mr. Elwood."

We moved on to the entrance to the Machine Shop, looked at the rows of lathes. I hoped I could see Mr. Gromberg, my Instructor in Tool Making and introduce him to them. He wasn't there. While I rattled on about Mooseheart's Vocational Educational Program, I could sense that we were moving in different worlds. I pointed up to the Sheet Metal Shop and ventured to talk about learning to solder two pieces of tin together. I was almost ready to give up all hope of getting a good tip for showing these guys Mooseheart.

I told them I wished I could have introduced them to my instructor in tool making. Mr. Gromberg says he can get me a job in the Elgin Watch Factory when I graduate here. But I don't think I want to work in a watch factory. I guess I don't know what I want. Maybe I can get an appointment to the Naval Academy or West Point or get an offer from some college to play football. Then I'd like to study law.

Mr. Droop put his arm around me and said "Son, we want you to do what you want to do. When you graduate you get in touch with us and we will find some way for you to get a job or go to college."

"Gee, sure, thanks, Mr. Droop," I said. But he gave me a kind of creepy feeling—hugging me and calling me son.

"We don't want to go in there" I told them when we

came to the one story laundry building. "It's hot and the machines make a lot of noise. Every Tuesday the guys in my hall send their clothes out with the bed sheets and things. We get our laundry back on Friday. Every kid has two laundry bags with the same number—my number is R-191 and it's on all my clothes."

Next we came to the two storied Hope Building. "There are four halls in that building: North Hope, South Hope, East Hope and West Hope. Forty guys can live in each hall," I informed them.

"After we first came here, I was put in the lower one closest to us, North Hope, for a couple of weeks"

"We call the six halls in the three-storied Industry Building up ahead the 'Industries.'" I explained that like the Hope halls, each one of the Industries has about forty guys living in it. I pointed to Upper South and told them that Cliff, Clyde and I lived there for about a year before we moved to Middle South to be with our buddy Dough Guinn

At Chapter Hall we turned right onto the road leading toward Mooseheart Lake. When we came to Mr. McCoy's farm office we turned left and drove out to the Dairy Barn. I pointed out the place where the dairy barn had been before Mr. Roselle had it jacked-up and moved to make space for the Boy's Village that the Moose wanted to build for us. [It was never built because of the depression.]

I thought they might be interested to know that the Mooseheart famous herd of black and white cows won prizes all over the world. My guests didn't light up on hearing about the prize herd, but they grinned when I complained that we were expected to drink a quart of milk every day. I ventured the opinion that it "Don't taste any different than the milk we got from our ordinary cows on Spruce Knob and in Richwood."

"Do you have to work in the dairy or on the farm over there?" Wesley wanted to know.

"Naw," I said and explained how Mr. Davis had to go to work in the steel mills when he was eleven. "He don't think kids ought to have to work except to learn a trade. Only the guys who want to be farmers work there to study their trade. Mr. Davis wants us to get a trade first. And then we can get a high school education. If we want to, we can study to go to college. We don't have to work on the farm or do nothing unless it's to earn money for ourselves."

We returned from the farm road to the road leading toward the lake and passed the new houses that had been built for Mooseheart officials like Mr. Bird, who was in charge of Education and Chaplain Payne our Protestant minister, and a couple of other Big Shots.

The man from Summersville, his name was Wesley something—I think it may have been Wesley Johns. He wanted to know if I would show them where I lived.

"Sure," I said, glad to know there was something they wanted to see besides the boys clothing store.

After passing the new houses for officials, we came to the row of five cottages on the right hand of the road. Quaker City, my hall, was the first in the row. Krebs Hall was the second. Both were "privilege halls" for Junior and Senior boys.

I led them into my hall and introduced them to Miss Pratt. She was a tall gray-haired lady, who looked like George Washington. We even called her George

Washington when she wasn't listening to us or telling us what to do. She showed the visitors from West Virginia through the hall and talked a leg off of them--worse than me. When we got outside, away from her, they asked me if she was mean to us.

"Naw, she's just fussy and kind of prissy, tries to boss us around, tries to get us listen to classical music. She's a 'mouldy' cook. Mrs. Jones in Krebs is a 'darby' cook--a real good cook. I'm going to try to move next door into Krebs next year, when the seniors graduate."

I showed them Walnut cottage where we stayed for two weeks in Detention when we first got here. Pointing to Elm Cottage, I told them about Bob Hanke's family: "Bob goes with Dough Guinn's sister Carolyn. Mrs. Hanke is one of the really Big Shots in the Women of the Moose. Next to the Hanke's house is Mr. McCoy's house, Maple Cottage. He's in charge of the dairy barn and all the Mooseheart farms."

They seemed to be bored and I was getting pretty certain that they wouldn't tip me much more than a quarter (if anything) for my effort to show and tell them about Mooseheart. It was getting late and I was as tired of them--as they seemed to be of me.

We buzzed around the Mooseheart Lake without stopping at the cemetery where no one that I knew well had been buried--except George Duffield. Camp Adams had been torn down. I didn't stop at Heart Springs or the place where I found that perfectly shaped arrowhead.

We came to the end of the circle around Mooseheart Lake where it spills over the road into Mill Creek.

"We're here at a corner of the girl's campus," I told Mr. Droop. "I don't know much about the girl's campus, but everybody that I take around Mooseheart really likes to see the Baby Village." (Photo 11) They said they would like to see the Baby Village. I was glad. I knew that if anything could soften them up to give me a good tip it would be the Baby Village.

The Pennsylvania Baby Village consisted of five identical buildings, four residence halls for babies and their nursery school. They were built on a scale to be smaller than any of the other residence halls at Mooseheart. The buildings in which the babies lived were named for four counties in Pennsylvania: Schuylkill, Susquehanna, Juanita, and Allegheny.

I let the two men to wander through one of the halls, and gave the housemother a chance to talk. The little boys gathered around two visitors for attention and affection. The men picked them up and held them and talked to them. No one I ever took to the Baby Village could resist the bright-eyed, neatly dressed, little kids. They bubbled while showing their toys and talked a mile a minute, not knowing they were working for me. The men from West Virginia melted in their charm.

Of course enough is enough, so I moved in for one last try for a good tip. I asked the matron to show the men the little boys' bedrooms and bathroom. She did and the effect was as usual: the miniature white thrones, less than half the size of an ordinary toilet, struck and resonated on a common chord of humanness: I half expected them to exclaim: "These little angels are people like us—just smaller. Isn't that just marvelous! "

I kept close track of my tips for guiding. And on the two occasions when I didn't show the Baby Village, my tips were in silver, not in paper money.

Photo -- 11 Wading Pool at Pennsylvania Baby Village

On the way back to the Campanile we passed the Philadelphia Memorial Hospital. I pointed it out but decided not to tell them about the time Dr. Nichols took out my tonsils, or set my dislocated arm by pulling it back into its socket. I didn't even try to tell them that Dr. Nichols played football for Butler University and was considered the greatest college half-back to play before Red Grange wore number 77 for the University of Illinois.

I wanted to get them on their way before the charm of the Baby Village wore off.

When we got to the Campanile they got out with me. Mr. Droop promised to write to me when he got back to Morgantown. He said I could get into West Virginia University on in-state tuition because I was really a resident of West Virginia. Then he shook my hand and said, "I'll try to help you get a job if you decide to go to the University."

Mr. Wesley thanked me for showing them around the campus, and wished me well in whatever I decided to do. Then he shook my hand and turned and got into the car and they drove off.

I stood there and watched them and waved until the car turned right onto the Lincoln Highway and disappeared. Then I reached into my pocket and pulled out the money they gave me when they shook my hand. Each man had each given me a five-dollar bill—twice the biggest tip I ever got before.

TRAINING

MR. GROMBERG HANDED each of the beginning students in the Tool Making class a block of metal about four inches long and an inch square. (Photo 12) The tall vinegary looking man, whose appearance contrasted sharply with his quiet precise kindness, explained: "this is a piece of tool steel."

Then he picked up and identified a coarse file, a fine file, a steel ruler, and a micrometer. He showed us how to use them. The micrometer measured objects to within one thousandth of an inch. When he finished those first explanations, he showed us the tool room where we could draw the tools we needed and assigned each of us to a vice on the long work bench.

"Now," the teacher continued, "I want each of you to shape, to file, this piece of tool steel into a block exactly four inches long and one inch square. Then bring it to me. I'll show you how to make your own hammer. If you have any questions get Earl Guinn or Al Beckwith or one of the older boys to help you. If they can't help you come to me."

That was early one afternoon in 1931. I was on my way; I fantasized, to prepare myself for a job in the Elgin watch factory. Mr. Gromberg promised that the company

Photo 12 --Tool Making Class in the Machine Shop

would be glad to hire any of us who completed his course. He had worked for the watch factory for years and knew that they wanted trained tool and die makers. He did not know that the depression would soon close the factory. Eventually Elgin watches would become collector's items. I was well prepared to major in tool making. I had elected and completed several of the compulsory pre-vocational training courses in earlier grades. Those courses included sheet metal, house painting, ornamental concrete, and Mr. Elwood's class in drafting.

I worked and studied at each of them for three months. In the tenth grade I elected to major in tool making. Throughout the calendar year (except for holidays and two weeks of summer vacation) our afternoons from 1 to 4 o'clock were devoted to the study and practice of "life skills," in art, music, and vocational trades—as distinct from academic study in reading, writing, and arithmetic.

James J. Davis and the people who directed Mooseheart were clear in what they meant by their policy of "training for life." When they talked about every student being "entitled to a high school education and a trade," they meant, every student must have a trade, and they are entitled to a high school education as well--if they want it and can complete it.

Dr. Verne A. Bird, Director of Education, writes in the Mooseheart Year-book and Annual for 1931:

> The aim is to take full account of the needs and capacities of the individual child and fit the instruction to that individual, to the end that he may best adjust to a complex, changing social and physical environment.

In the opening lines of his "Preface" to The Iron Puddler James J. Davis recounts a story about a boy who was in danger of being thrown into jail in Elwood, Indiana, following his arrest in a freight yard where a box car had been robbed.

"Where were you previous to the eighth and immediately subsequent thereto?" asked the city attorney.

The boy could not understand the question that the attorney hammered at him three times. A kindly City Clerk [James J. Davis] translated the question for the boy. The boy was able to explain where he was when the boxcar was robbed, and escaped going to jail.

Davis's reaction to the Elwood incident leads him to assert "this long necked jargon must go. It is not the people's dish." Davis continues, explaining my training and schooling:

> When I was planning a great school [Mooseheart] for the education of orphans, some of my associates said: 'let us teach them to be pedagogues.' No, I said let us teach them the trades. A boy with a trade can do things. A theorist can say things. Things done with the hands are wealth, things said with the mouth are words. When the housing shortage is over and we find the nation suffering from a shortage of words, we will close the classes in carpentry and open a class in oratory.

It is significant that Matthew P. Adams, the Superintendent at Mooseheart from 1915 to 1927, and Ernest N. Roselle, his successor from 1927 to 1935, were both nationally recognized as leaders in vocational education. They were not just trained teachers; they were scholars and leaders in a branch of education that dominated schooling at Mooseheart while I was there from 1926 to 1933. Girls and boys were trained to work with their hands in about twelve pre—vocational courses ranging from Agriculture and Art to Woodworking. Generally speaking a boy or girl could get training in any recognized trade while they were at Mooseheart. And if they finished High School before 18 they could take post graduate high school vocational courses.

Moreover, Mooseheart also provided several pre-professional courses including pre-law, pre-medicine and pre-dentistry. For instance, I elected to stay an extra year to prepare for the examinations for the Naval Academy and for West Point. And then, I left Mooseheart before classes for the new school year began.

The choices and opportunities given to me by the Vocational Education Program have been a truly beneficial force upon my personality and my life. The vocational program was flexible enough to permit me to shift from tool-making to a college preparatory major and take classes in typewriting.

The college preparatory major enabled me to speed through my AB degree at the University of West Virginia. Then in 1942 that degree qualified me for a commission in the United States Navy. My military service enabled me to earn a Master's degree in Dramatic Art under the GI bill. The Master's degree helped me to find a job as Technical Director of the University of Kentucky's Guignol Theater. My vocational training in drafting, painting, sheet metal, and toolmaking equipped me with the necessary manual skills and confidence to design, build, paint, and light theater scenery for eight years while I earned my PhD degree in English Literature—specializing in the history of the Shakespearean Stage. I owe much to the manual skills I developed through James J. Davis's vocational training program at Mooseheart.

LEARNING

THE NEW HISTORY teacher was hired to coach the basketball team. He came into our class, called the roll, picked up a piece of chalk and wrote on the black-board: "George Washington was not the first president of the United States." I raised my hand; Carlos Powelson acknowledged the signal.

"All right?"

"Why?" I asked.

"Go to the library and find out!" he ordered.

Then turning to the room full of students he said firmly: "All of you go to the library and find out why George Washington was not the first president of the United States. Class dismissed!"

I would like to report I discovered that John Hanson of Maryland was the first president of the United States—under the Articles of Confederation. I didn't, nor did anyone else in the class. But something important happened to me.

Coach Powelson told Jennie Hance, our wonderful Civics Teacher, "that Rhodes kid in my History class is a bright one—he's quick to ask questions."

Miss Hance told me what the coach had said. She had already told me that I was bright, because I chose to campaign for Norman Thomas in our mock presidential election. I had read about the Socialist candidate in The National Scholastic. But Miss Jennie told any of her students they were bright if they would read what she handed them to read.

If Coach Powelson told Jennie Hance I was bright, then I had to live up to it, for the coach of the basketball team was a figure to be revered. Besides, I believed he was a good teacher—especially since he thought I was bright.

My feelings about Coach Powelson, however, soon moved from reverence to doubt about his judgment—and especially his fairness. He refused to let me try out for basketball.

"I can't use you—you've got two left feet." Obviously he thought I was too awkward to play basketball.

How could a man be so acute in his judgment about my intelligence, and yet be faulty in his judgment of my potential as a basketball player? Perhaps he was no more astute about my intelligence than he was about my ability to run around in short pants and throw a ball through a hole.

Despite his probable accuracy about my awkwardness, my resentment was real and it persisted. Even so, Coach Powelson and I became friends. I had no reason to "suck-up" to him in order to play basketball. Instead I could talk to him, listen to him, argue with him, and when I thought I was sure of my ground, snap back at him.

Coach Powelson pointed me to the library in Aurora, and directed me in my early reading: Lincoln Stephen's Autobiography, and Clarence Darrow's The Story of My

Life were the first two books that I remember reading at his suggestion.

Obviously, I believe he was a better teacher than a basketball coach, although his teams frequently beat some of the very best high school teams in Northern Illinois and Northern Indiana in the early 1930's.

Earl Guinn played basketball for Carlos Powelson and idolized him. Years after leaving Mooseheart, I became a teacher. Throughout my career I paid him the most sincere of all compliments. I regularly tried to use his teaching technique of "shocking and surprising" students into learning, and encouraging them by trying to get them to read books. I have often, as he did, told "ordinary" students they were bright to get the best out of them.

However, my resentment lingers toward Coach Powelson for his refusal to allow me to try out for basketball. In the first place, he violated the basic tenet of Mooseheart's traditional athletic policy. Ben Oswalt, the first football coach at Mooseheart, established the practice of permitting every kid who wanted to play in any position on a team in any sport to continue on the squad as long as he turned up for practice.

Johnny Williams, bless him, kept me on his Third Team for three seasons as a left end. I played football in all of the practice scrimmages, as a substitute for other substitutes.

I started one game for Johnny Williams against Oswego High School. I even made a touchdown. Oswego was the school from which Slade Cutter went to the Naval Academy. He played football there, and he became a World War II hero as the Commander of a submarine.

If he had been a couple of years younger, I might have played against him "man to man." Ends at Mooseheart were assigned to block the tackles on offense. No question but he would have "whipped my ass" for Slade Cutter became All-American tackle at the Naval Academy.

Sports were an important part of our schooling, in the sense that the Greeks emphasized sports, and that John Milton discusses them in his essay "On Education."

Mooseheart had many other excellent teachers who helped me. Mr. Gocher was as good as the best. He taught us about the Periodic Chart and explained how a screw was really an inclined plane, and talked about Boyle's Law. He was a quiet spoken gentleman, in every sense of the word, and I despair to this day that I can never measure up to him.

I was at the blackboard in Miss Crane's class diagramming a sentence in English when someone from the principal's office opened the door and walked in. Something radically different had been happening in American education—and it reached Mooseheart in early 1927. After looking briefly at a piece of paper handed to her, Miss Crane began calling out the names of about half of us in the room.

We were shepherded in an unorganized group to Mrs. Grave's class next to the end of the second row of portables—the temporary wooden buildings used as classrooms. In the midst of diagramming a sentence, grammar was wiped out for me and for those in my group. We missed that approach to understanding and using our language. I was never again in grade school, in high school, or in college exposed to grammar as tool in writing. Because of my later training (or lack of training)

I have been unhappy about the decision to drop the study to grammar at Mooseheart. However, I believe Mooseheart's approach to schooling was generally both realistic and enlightened.

In Mrs. Graves's class we read the Bobbsey Twins books and similar stories. I don't remember a single incident about the twins. I do remember the class was usually boring. We also read long articles about how sheep were led by a black goat to be hit in the head, skinned, and turned into lamb chops—or perhaps it was pigs turned into pork chops. As I recall, we were told at the end of the article that "nothing was wasted but the bleat of the sheep," or maybe it was the "squeal of the pig."

One of the memorable teachers in my grade school classes was Miss Stuart. She would show up in Mrs. Graves's class twice a week, play music on a phonograph records and talk about the piece. She went through about twenty standard works. I still remember: The Anvil Chorus, The Pilgrims Chorus, William Tell Overture. The Soldiers Chorus, To a Wild Rose, and To a Water Lily, Pomp and Circumstance, and Rose Marie from a popular light opera. We were drilled and tested on how well we could recognize the several works. Those pieces mark the beginning and alas the end of my training in music. It wasn't Mooseheart's fault—it was my decision to play football instead.

Training in music and performance was stressed at Mooseheart. I tried to learn to play a tuba because Dough Guinn carried one around in the marching band and apparently played with skill. After several weeks with Mr. Warrick on the tuba—I quit. He did not encourage me to give up my practice or to suggest that I ask for lessons on another kind of musical instrument—I just was not interested in music—I quit. And I regularly skipped the compulsory practice for the massed Chorus singing Handel's Messiah.

Beginning in 1931, while I lived in Quaker City and Krebs Hall, I skipped many of the Sunday afternoon concerts. Those Halls did not have a Procter, only a House Mother; she never inspected the bathroom on Sunday. So the bathroom became the library where I did much of my reading.

From three to four o'clock on Sunday afternoons, I could usually be found with a book in my library. Why not read? The alternative was to listen to The Philharmonic Orchestra, or to Flossie Ford play the Organ, or to one of the three or four Glee Clubs, or hear, again, the All High School Mixed Chorus of 350 voices present a cantata about building a ship.

It was always a mystery to me why the music staff so seldom had the marching band in an All Sousa concert. And as far as I remember they never did allow our dance orchestra, the Davidsonians, to perform in concert. That dance band could really play such pieces as Ninety Nine out of One Hundred, Mood Indigo, Stardust and Home Sweet Home. Serious music was not my special interest. I did attend concerts occasionally, with Virginia Generis. But I preferred books to music.

Schooling at Mooseheart could also happen outside the classroom. Mr. Wahl the math teacher taught me something about older men—or good manners—or both when I was in the ninth grade. It happened when he picked me up after class and offered me a ride to Aurora. I had a permit to go to the library in Aurora and return a copy of Gay Neck, the story of a heroic carrier pigeon that saved a lot of men during the War in France. Anyway, he stopped his car and picked up me and Miss Jewell, the new math teacher. I jumped in the car and plopped down beside him.

"Young man," he growled, jesting--but irritated, "you are old enough to know how to share the wealth—you ride on the outside so we can both enjoy the company of this young lady."

I got out and let Miss Jewell in beside Mr. Wahl and then climbed back into the car. I discovered that apparently middle aged men like pretty young women. They talked and laughed and generally ignored me. He stopped at the Library, let me out, and drove away with Miss Jewell—she was really a jewel.

Mr. Wahl also introduced me to algebra, suffered me through advanced algebra, took me into geometry and pushed me through solid geometry and plane geometry. There I learned the difference between "Plain" and "Plane."

Mr. Wahl, Carlos Powelson, Mr. Gocher, Jennie Hance, and other teachers like Debora Hart, dramatics; Zola Kinney, Biology and Botany; Miss McDowell, Miss Crane, and Rhoda Houseman did their best to shape me academically through my years in grade and high school.

Still, as I have said before, the emphasis at Mooseheart was not on Academics but on Vocational Training. Mooseheart officials and teachers talked regularly about "Training for Life." They never forgot that all Mooseheart students were to be taught a trade. We were all to be prepared to earn a living working with our hands.

WORSHIPING

JAMES J. DAVIS, with the others who helped establish Mooseheart, decided that every child should be brought up in the faith of his father. Since our family was Baptists, we worshiped on Sunday with the other Protestants. Catholics made up about half of the students at Mooseheart. We were required to study our Bible or Catechism for fifteen minutes every night five nights a week—Sunday through Thursday.

Our Protestant Minister, the Reverend C. Donzel Payne, served in France in World War I and was addressed as Chaplain. He was a quiet retiring man compared with Father Laffey, the Catholic Priest, who was enthusiastic, and a strong political and social force on Campus.

Chaplain Payne drew on his military experience and upon his studies in Chemistry and Physics to illustrate his sermons. He frequently used an analogy, to persuade us towards his faith: "There is no taste in the white of an egg without salt--just as there is no taste in life without the salt of religion—without a belief in Jesus Christ." I recall in one of his sermons that he assured us there is enough energy in a snowflake to destroy the world. He then pointed out that God created the world and all of the stars we see and those we cannot see. "Consider it," he urged. "Consider the power God has to create the universe, and if He wills to destroy it. Consider the might of his power with that of a snow flake that has the power to destroy the world."

The Chaplain held Class Room Religious Instruction for Protestant Children while I was at Mooseheart. And I recall for a time zealous white-haired old man from Batavia was drafted to instruct our Sunday school class of teen-aged boys. He was scary in his exhortations to: "surrender your souls to Jesus or burn in Hell's fire forever." He would continue: "bathe yourselves in the blood of the Lamb if of you want to live forever."

One time the Evangelist Billy Sunday came to Mooseheart to preach to us. He brought considerable credibility. Because he had played major league baseball—-I think for the Chicago White Sox.

Billy lined up ten different-sized and different-colored glass vases across the stage on which the pulpit stood at one end of the Assembly Hall where we worshiped on Sunday. I was particularly grabbed by the beauty, the shape and scarlet hue of the one of the vases that shimmered in the auditorium light.

Billy told us the story of a little boy like us whose parents gave him five cents to put in the collection plate and sent him off to church. The boy saw a bunch of "bad boys" playing marbles "for keeps" on that bright Sunday morning—-"gambling on the Lord's Day, mind you."

The little boy stopped, traded his nickel for five marbles and lost them. When he got home, his parents asked him if he put his nickel in the collection plate.

"Sure," the little boy lied.

"Now," Billy Sunday snarled with anger in his face, "What did that boy do?

"Let me tell you," Billy roared! He picked up a hammer, walked over and picked up the first of the beautiful colored vases, lifted it, smashed it to pieces. The colored glass fell and rolled toward the feet of the kids in the front row.

"That boy broke the one of the ten commandments of the Lord: honor thy father and thy mother," he explained.

The Evangelist picked up another of the vases and smashed it: "Remember the Sabbath day, to keep it holy."

Next, he roared, "Thou shalt not steal. That boy stole the money his parents gave to him to give for the work of the Lord. It was the Holy Lord's money," he shouted. The preacher then picked up another colored vase and smashed it. He was getting close to my especially beautiful scarlet vase, and I hoped he wouldn't select it.

The Preacher moved through the line of vases, smashing them one at a time, after charging the boy with breaking another commandment. Every time colored glass fell to the floor, I cringed—-I was on Peanut Burns' special labor gang that had to sweep up the glass after the services were over.

When the Evangelist finished he had the boy on nine of ten counts. I checked out the Ten Commandments in my King James that night during bible study. I concluded the only commandment the little boy didn't break was the seventh.

I had a vague idea of what the seventh commandment meant—some of the big boys used to joke: "I wonder if infants have as much fun in infancy as adults have in adultery?" Incidentally, the surviving the vase was the scarlet red one that had caught my particular interest.

I used the story of Billy Sunday's performance in a drama class while I was an Instructor working on a PhD at the University of Kentucky. One of the students was an Evangelist trained at Wilmore College near Lexington. After class he came up to me and grinned:

> I heard that story before. Actually, I tried to use it once at a revival in Covington, near Cincinnati. I went out to a dish barn, he reported, and bought ten different colored vases. Then I stopped by a hardware store and bought a cheap hammer. I lined the vases up and began my sermon. When I came to the place to break the first vase, I walked over with my hammer, picked up a vase, and hit it and--The hammer broke.

We attended church services at Mooseheart every Sunday, and put a check in the collection basket for five cents to be drawn from our Student Bank accounts. I do not recall to whom the checks were made payable. But I kept close watch on my account. The Mooseheart Protestant church never took a penny from my account. Many of the kids I knew didn't have any money in the student bank so their checks should have "bounced." I concluded that the Loyal Order of Moose provided its kids with a place for worship and tithed for us. Certainly Mooseheart offered all of us an opportunity to be solidly grounded in the faith of our father.

The use of the of Assembly hall for both Catholic and Protestant worship services was constructive in shaping the cultural and spiritual development of every Mooseheart child. We learned that religion made absolutely no difference between the kids we lived with.

After I left, Mooseheart built an imposing House of God. It replaced the assembly hall as a place for both Catholic and Protestant worship services. I hope the practices in the new House of God have continued the same sense of religious tolerance, and appreciation for religious diversity, that was offered to those who attended religious services in the assembly hall.

Blanche Rhodes' children entered Mooseheart with little if any tolerance for religious diversity—for diversity of any kind. We were Baptists, living at 16 Riverside Drive, at the junction of the road leading up the hill towards Greenbrier County within sight of St Mary's Catholic Church. Our elders (not Blanche or Carrie) told us, possibly in jest, that the priests and nuns might eat us. We almost believed that those black robed people, especially the women of St Mary's, were every bit as dangerous to little children as Gypsies. And we were almost certain that those bronzed ragged people with their litters of scrawny dirty little brats drove their wagons up the Greenbrier road looking for nice fat little white children to steal—presumably to cook and eat. No question about it, 16 Riverside Drive was a dangerous neighborhood.

During my final year at Mooseheart, as I reported earlier, one of my most positive experiences occurred when I sneaked out of Quaker City Hall after "lights out" to carry a sick puppy to Father Laffey for help. I not only expected his help—I got it. And I knew he would not report me for breaking rules. Even though I was not a Catholic, I was one of his Mooseheart kids

DANCING AND ROMANCING

ABOUT THE TIME Mooseheart boys began wearing long pants and the girls were fitted with bras, many of them began to pair off—as we said, "Goin' Together." Couples often began with a note that read like my first romantic effort:

Dear Ginny,

> Meg told me that you and Harry have quit goin' together. I told her that I like you a lot ever since I set behind you in Miss Graves class and pulled your hair. I remember you got pretty mad at me. And you said "quit it, creep." Ha! Ha!

> I don't think I'm a creep any more. Do you? I thought we danced good together last Friday. It was real swell. Don't you think? I only stepped on your feet a couple of times. Ha! Ha!

> Anyway, I'd like it if you would be my girl and we could go together. Send me your answer by Meg. If it's YES, I'll wait

for you at the grocery store tomorrow before school. You could sure use me to carry your Cello. Ha! Ha!

I hope you say YES because I like you so much and think you are sure pretty.

I love you,
Ernie Rhodes

P.S. I can't meet you after school today if it's YES because I got football practice. But I hope it's YES

Meg delivered Ginny Generis' answer after history class.

Dear Ernie:

My answer is YES. Wait for me after class by the girls stairs. I got a lot of books to take back to the hall. Ha! Ha!

Don't you ever ever pull my hair again— You NICE Creep. Ha! Ha!

Love and Kisses,
Ginny

I was at the bottom of the girls' stairs as soon as I could get there after the first noon whistle. Ginny and Meg got there finally, carrying about half the books in the library. The girls handed me the books.

"I'll see you guys," Meg said, and left.

Ginny told me she was working on a music project about the life of Bach.

I said something, but I don't remember what. I found it hard to think of anything to say to my first girl the first time we were going together. So I just floated along with a ton of books, answering her questions and grinning.

Ginny chattered about her project. She was real smart—probably the smartest girl in our class—and next to her sister, Prudence, the best looking.

Meg showed up from someplace when we got to the edge of the Girls' Campus and I gave the books back to the girls.

"I'll see you tomorrow and I hope you'll write to me," Ginny said.

Couples going together usually wrote love notes to each other. I realized that I wouldn't be able to finish Jack London's White Fang—not tonight, anyway.

After we parted, we were supposed to wave to each other two or three times. I waved until Ginny reached her hall. She stopped and waited until she was sure I could see her. Then she waved, turned, and disappeared around the corner of the building.

Because I had machine shop and football practice we did not see each other until before the first whistle for Bible Study. Ginny was alone at the corner of her hall when I recognized her in her blue dress. We waved to each other until the second whistle blew for Bible Study.

After Bible Study I wrote Ginny a love letter. I told her how much I liked her. I said she was the prettiest girl at Mooseheart (and it was true—except for maybe her sister Prudence who was taller and just a little bit better looking). Of course, I didn't put that in the note. I told Ginny how smart she was to be able to play a cello—and I hoped she wouldn't think I was dumb just because I couldn't read a note of music. I wrote about how happy I would be to carry the cello for her. I said I was jealous of it because she spent more time with it and was always closer to her than I could get. When Ginny played the cello she sort of hugged it between her legs, curving her arm around it to finger the strings as she moved the bow across them with her other arm. Could she understand what I was thinking? I hoped so. At the same time I wanted her to believe I was too timid to put what I was thinking on paper.

Shifting gears, I told her I respected her too much to want her to think I was rushing her if I asked: "Can I kiss you after the dance tomorrow night—if we get a chance?"

I carried her cello to school Friday morning and we exchanged love notes. I was encouraged by what she wrote. She said that she was glad that I had asked to go with her, that she thought I was a nifty guy and a good dancer. She wrote that two girls in her hall told her that they thought I was "cute."

Although Ginny hadn't read my note, her note answered my question about kissing her after the dance. She said she hoped I wouldn't try to kiss her after the dance, but if we got a chance to smooch, she would not be too mad at me if I tried—so long as we didn't get caught.

My sister, Dorothy, told me that the advantage of going steady was having someone to dance with besides her brothers and their friends. For me, the advantage of going steady with Ginny was having one of the prettiest girls in our class for the Home Sweet Home dance. After the dance I would have a chance to get in a "smooch" as we left the auditorium and walked down the steps into a dark spot. The two lights were never very bright on the ornamental concrete poles in the low stretch between the bottom of the auditorium steps and the steep flight leading up the sidewalk. And if it was dark enough, I might even get to kiss her twice. Then there was still a chance for another kiss on the walk leading to the place where it turned off to the girls' campus. I always think of this stretch of sidewalk as "Smoochers' Lane." Getting a smooch depended on luck in finding a place behind a couple of Mooseheart Mothers. We had to keep away from chaperones like Old Hatchet Face and Miss Gayle, the Supervisor of Girls. At least I would get to hold Ginny's hand as we walked from the auditorium to the sidewalk leading to the Girls' Campus.

It was difficult to understand how some guys in the Davidsonians Dance Band could sit on a chair and toot a horn or beat on a drum—when there were girls like Ginny Generis to hold.

During the first dance of the evening we all watched to see which new couples had been formed and which couples had broken up. It was time for guys who weren't going with someone to move when the boy of a known couple headed across the dance floor for a new partner. And we also watched the former girl friend waiting for some new guy to move in and ask: "May I have this dance?"

After the first few dances with Ginny and the obligatory turns with my mother and her friends and Dorothy and Meg and a couple of my friends who had no guy, I was free to try to keep Ginny to myself. It wasn't easy. Ginny had a lot of friends and there were half a dozen guys who would try to "scab me out" if they could get up the nerve to try. They usually had plenty of nerve.

The chaperone's practice at all the dances was to keep the couples "goin' together" apart and dancing with others—especially the wall flowers. Miss Gale shook her head with disapproval when we danced together too many times and I tried to hold Ginny too closely. Guys I didn't like would appear from nowhere and ask Ginny to dance. She was expected to say, "Yes." The chaperones tried everything from "tag" dances to "Girls' Choice" to mix the steady partners up.

Also I enjoyed Friday night dances because they gave me the opportunity to dance with two or three girls whose softness fitted closely to me, and whose hair smelled nice. One of them was Ginny's sister Prudence.

They dimmed the lights slightly when the band started Home Sweet Home. I walked across the floor and claimed Ginny from some big freshman crumb that had been trying to horn in.

We floated over the floor. Ginny whispered, "We're nice together, don't you think?" I answered: "I think," and held her tighter.

When Home Sweet Home ended we maneuvered our way into the crowd in front of Mrs. Van Sickle and Mrs. Davidge—two of my mother's close friends. I knew they wouldn't snitch if they saw us kissing.

I remember getting out of the auditorium and down the steps onto the lawn between the auditorium steps and those leading up to the sidewalk. It seemed to me to be dark enough; so I turned my head and tried to kiss Ginny.—I missed her lips and kissed her nose. She drew back startled rather than enchanted. I opened my eyes frustrated and puzzled—a nose is not a pair of soft lips.

It was dark, but it was light enough for me to look directly at Old Hatchet Face, towering above the relatively two smaller friends of my mother.

Ernest Rhodes and Virginia Generis headed the notice with three other couples on the NO COMPANY list distributed Monday afternoon to every hall on campus. NO COMPANY meant two weeks during which we would miss two Friday Night Dances, two Saturday Night Movies, and were forbidden to associate with each other.

Our romance survived two weeks of NO COMPANY. Then It limped rather than skipped along for about a month. Then a guy I already disliked began to horn in.

Meg delivered the note from Ginny "canning" me. Meg and I had always been kind of buddies—confidants. We began leaning on each other after Meg "canned" the guy she was going with. Meeting by accident, more or less, we sometimes walked together to school—casually holding hands.

I made a point of rescuing her from the wall at dances. Like Ginny we fitted together when we danced as if we were made for each other. Meg was sort of pretty—cute and sparkling, but not really beautiful like Prudence Generis.

One Friday night during the dance, Meg and I sneaked out onto the fire escape and started smooching. We danced together and "smooched" twice—without getting caught—between the steps of the auditorium and the steps leading up to the sidewalk. Meg helped me kiss her lips rather than her nose.

We started "goin' together" without asking, or waving, or writing love notes. We worked out a system with another couple, Willie and Rose, of using the stairs at the High School as a place for "smooching." We called our system "Meeting in the Library." We actually met in the Library and checked out books or turned them in. Then we would straggle out of the Library to the stairs.

Meg would leave first and go the bottom of the stairs to watch if anyone came along. Willie and Rose would follow Meg part way down the stairs and stay there and "smooch," while I went to the top of the stairs and stayed there to warn them if anyone came along. The next time we held our "Meeting in the Library," Meg and I got to use the stairs for smooching while Rose and Willie served as sentinels.

Meg and I became an informal pair sharing our secrets and the pleasures of dancing and smooching. We were together longer than we were with anyone else. It ended just before my 18th birthday with a question: "You still got a crush on Prudence Generis?"

"Lot of good it does me, as long as that creep is around."

"Willie and Rose have quit," Meg told me.

"Well?"

"Willie wants to know if I'll go with him."

"Willie?"

"Rose quit him because she says he's got a crush on me."

"You got a crush on Willie?"

"I guess I've got more of a crush on him than you have on me," Meg said.

Well that was that—my two best friends started goin' together. But it gave me a chance to ask Prudence Generis if she'd be my girl.

I was usually Blanche Rhodes' "biddable boy." Some might say that I was a "Mommy's boy." That suspicion may explain what happened at a Friday night dance shortly after my 18th birthday in April 1932.

I had danced the first dance with my mother. I usually danced the first dance with Meg. But she was at that very moment on the floor with Willie. I was confused and irritated—actually rather hurt, when mother asked me "Why don't you dance with Helen?"

Helen was a skinny girl with glasses, the daughter of one of mother's friends. And the question took the tone of one of my mother's implied commands.

"I'll dance with who I damned please," I snapped.

Mother teared up and I knew that I had hurt her. When I recall the incident, I cringe, and feel remorse. But it had to happen. I can smile when I reason it out. Most boys I have known declared their independence from their mother years earlier than I did. I had indeed been a dutiful son, but on that night I wasn't inclined to honor my Dear Mommy's implied command.

I left her moist-eyed and shocked. Squaring my shoulders, I turned and walked across the dance floor to make my move on Prudence Generis, the girl I was sure I loved.

STRIVING

"HEY RHODES," COACH Seeglitz called.

"Here, Coach," I answered.

"Warm up!" he ordered.

I was finally getting into my first varsity football game after three years on the Third Team and a year on Coach Gocher's Second Team.

What a chance!

I was sure my hour of glory had arrived early in September 1932, near the end of the final quarter of the opening game of the season. We were in Richmond, Indiana, playing before leaders of the Indiana State Moose Lodges, gathered to see "their boys" whip the Valparaiso University football team.

Lifting my knees to my chest, I pranced down the sideline for instructions. Consider the situation: the game was scheduled with the understanding that Valparaiso would not use its players weighing over 200 pounds. Mooseheart was winning at the half 6 to 0. At the start of the second half the Valparaiso coach realized he couldn't afford to lose to a high school and put in his better and heavier players who pushed ahead to lead 19 to 6. Coach Seeglitz was calling on me with less than five minutes to play.

Valparaiso punted. The ball sailed down the field towards our quarterback as I pranced up to the coach, knees to my chest, headgear in hand, ready to rip. I knew what my instructions would be: "Tell Willie to get a pass to you behind their right halfback—he's been creeping up all day to stop our runs."

Coach Seeglitz didn't get a chance to deliver his instructions as their quarterback fumbled Willie's punt and we recovered it on the 35 yard line. Coach was too excited. When he finally noticed me he ordered me to "Keep warmed up!"

I turned and pranced back down the sidelines lifting my legs almost up to my chest. The business of high-stepping can wear a guy out. But I knew I had better keep warm and loose; I would need all my speed to get down behind that halfback. Boy, I knew I could do it, if Willie could just get the ball to me. There was still time for me to score a touchdown and make an extra point on 99-5 which called for a short jump pass to me just over the line of scrimmage.

At 19 to 13, all I had to do was get down the field on the kick-off, force a fumble with a bruising open field tackle, and recover the ball. With two or three minutes to play, there would still be time for another short 99-5 to me and a short dash for a touchdown to tie the score at 19-19.

Then Willie could cross Valparaiso up with a 92-2, a jump pass behind the line to the right end, for the extra point to win the game for Mooseheart 20 to 19. Certainly, Willie should try the 92-2 pass to the right end for the tie-breaker—because Valparaiso would be keying on me. And "after all," I reasoned, "Mooseheart football is a team game. There is really no place on it for stars."

As dreams of glory surged through my head, and the clock ticked down towards the end of the game, Mooseheart held the big Valparaiso fullback.

The ball came over to us on downs, deep in our own territory. There still time for Willie to complete a long pass to me for a touchdown,

I was still warming-up, high stepping and dreaming of glory, when Coach Seeglitz remembered that he needed me.

"Rhodes," he growled.

"Yes Coach."

Before the Coach could speak, one of our halfbacks fumbled—and then recovered the ball.

By the time the coach remembered he had called for me. He hesitated, searched and found a use for me: "Got your headgear?" he asked.

"Yes sir, Coach."

"Well hold onto it," he ordered, "and collect any others you find down at that end of the bench—they cost thirteen dollars apiece."

I earned a varsity football letter, by appearing in enough games to average a quarter for each game we played during 1932. I forget if that letter found a place on a sweater warming the heart of my friend Meg. Probably it was on a sweater worn by Prudence Generis—whose heart really couldn't be warmed—by me.

My participation in football would end with the 1932 season and a single play in spring practice for the 1933 season.

Before my graduation in June 1933, I petitioned to

stay at Mooseheart for another year. I needed additional classes in mathematics to prepare for the competitive examinations for West Point or Annapolis. It really didn't matter whether I became a soldier or a sailor. I wanted to go to college and to play football. I reasoned that if I stayed at Mooseheart for an extra year, I might be able to get an appointment to one of the Military Schools. And with another year of high school football, I could put on enough weight and be seasoned enough to make a football team at one of the Academies.

With high hopes, I turned out for spring football practice in 1933 and a second year on the Mooseheart varsity. Herb Heilman I were the only returning ends who were lettermen from the 1932 varsity. We had clear shots at the two starting positions. Our only competition would be from our younger brothers, Cliff Rhodes and Jim Heilman. Both would be moving up from the Second Team.

Spring practice ended with the traditional game between the graduating seniors of 1932 and the tentative varsity for 1933. Rhodes started at left end for the 1933 varsity. That is, Cliff Rhodes started. My younger brother "scabbed" me out of my position on the team.

I got into the game for one play, the final play of the first half. My assignment was to block downfield. As the play developed I saw I couldn't get either the fullback or halfback. So I went for the safety—the quarterback in the diamond defense that we used at Mooseheart in the spring of 1933.

The quarterback was one of the few guys at Mooseheart I never liked —a fancy dancey guy. He only threw one pass at me while we were on the varsity. It was too high so I bobbled it and missed it. Any way "Snotty" feinted to his left and hopped back to his right. I was slow in reacting to his feint when I threw myself at him. It was probably an accident, but I slammed into him and hooked my right leg behind his left and held it there while I twisted my body. I stretched out both my legs and arms as he crumpled and landed flat on his back—with me on top of him. Snotty started squawking:"Holding! Holding!"

The referee didn't see me hooking and didn't call me for it. Pleased with myself, I lay there across Snotty's belly with my arms stretched out farther—just to show I wasn't holding him. It was my most satisfying moment in five years of striving to be a Mooseheart football player.

During his half-time criticism of the varsity team's play, Coach Seeglitz turned to my brother Cliff and said: "Cliff, that downfield block you made on the last play— that's the way to do it! --- that's the way to do it, boy!"

I didn't open my mouth. And no one corrected the Coach, although I noticed a couple of guys grinned at his mistake. That incident ended football for me.

I had only one real triumph in the Mooseheart athletic program. A few days after the football game between the varsity and graduating seniors, I showed up to draw equipment and try out for the varsity track team.

Coach Seeglitz shook his head at me and said "Rhodes, you're not interested in track. And you're not fast enough. All you want is a jock strap."It was a generally suspected most Mooseheart guys considered their athletic supporters as necessary equipment for Friday night dances: to keep their maleness under control.

I decided to take part in the annual track meet which

was open to all junior and senior students including the varsity. Nunzio Ferrara and I were both seniors and roommates at Krebs. He wanted to improve his endurance while I wanted to improve my speed. So we worked out a training schedule and followed it for about a month. Every morning we sneaked out of Krebs before the 6:30 whistle and ran around Mooseheart Lake finishing the last part of the trip by running at full speed.

By the time the meet was to take place; I believed I could beat any of the varsity's quarter milers except Louie Pastor. When Louie decided stick to the 100 and the 220 yard dashes, I was confident I could win the quarter mile. Coach Seeglitz teased me when I showed up at the equipment room to draw some hand-me-down spiked shoes for the meet: "I hear you're going to beat my quarter milers."

"That's right," I said. "And if I do will you let me on the track team?"

"Sure, if you can beat them."

"Will you issue me brand new equipment— a jock strap, trunks, jersey, socks, new spikes, and one the new red sweat suits?"

"If you win, you can have them," he promised—right in front of all the kids in the issue room.

I beat all of his quarter milers! I was lucky, Louie Pastor wasn't in the race.

Coach Seeglitz kept his word.

That win in the quarter mile (440 yard dash) won me a place on the 1933 varsity track team. And I was issued brand new equipment. I finished second with a 2:08 half-mile at St. Johns Military Academy in Wisconsin. Later in the Triangular Meet between Culver, St John's, and

Mooseheart I ran two 440's both under 56 seconds and on the same day—one on the Mile Relay team the other on the Medley Relay team. That was probably my best ever performance as a Mooseheart athlete.

As a freshman at West Virginia University, I tried out for the West track team and won the half mile (880 yards) with a miserably slow 2:13 on a five lap indoor track. During practice one afternoon, Art Smith the tall, impatient, and mean- tempered coach of the track team kicked me-- literally kicked me in the ass-- for something I did or didn't do. So I quit the WVU track team. I really didn't have time for it. I had to work to eat while I tried to keep up with my studies. And track is not football.

After many years, I guess that Carlos Powelson and Coach Seeglitz were right. I suppose, I was too slow and clumsy to be a good athlete. But I cannot forgive either of them for refusing me the chance to do as well as I believed I could do.

Mooseheart's athletic program shaped the lives and characters of all four of Blanche Rhodes' sons. We all played varsity football and took part in track. Cliff and Earl also played baseball and basketball. Earl was elected captain of the junior varsity basketball team, and then the next year earned his letter playing on the varsity. Blanche's youngest son Clyde "Babe" Rhodes became one of Mooseheart's greatest quarterbacks. He earned a football scholarship to support his studies for a college degree from Lawrence College in Appleton, Wisconsin.

Earl and Clyde were elected to Mooseheart's Athletic Hall of Fame for their all-around athletic achievements. Cliff and I had to be satisfied with lesser contributions to team sports, developing "strong bodies, sound minds, and pure hearts." Cliff spent much of his time coaching young kids' baseball teams in Santa Barbara. I often watch NCAA football on TV, and I attend every basketball game I can when the Old Dominion University women play.

LEAVING

AFTER MY CLASS graduated in 1933, I stayed on to study mathematics and prepare for entrance examinations to Annapolis and West Point. I was lonesome and unhappy. I wanted something, but wasn't sure what it was. Everything seemed to go wrong.

I suppose one of the causes for my discontent was that I hadn't found a girl who interested me. My romance with Prudence never flourished. Almost the same thing that happened with Ginny also happened with Prudence. We got caught smooching in a closet during a party in Quaker City. For Prudence the punishment was especially severe. We ended up on NO COMPANY for 30 days and she missed her graduation dance. She was furious and canned me. The truth is she was a real nag even if she was the best looking girl at Mooseheart. I was glad when we broke up. I didn't miss either of the Generis sisters when they graduated. But I sure missed my buddy Meg. I expected her to write to me after she left, but she didn't.

I was drafted into the job of Secretary of the Mooseheart Lodge of the Junior Order of Moose. It gave me a kind of status and campus political power that I did not find satisfying.

The JOOM decided to stage a play called, Girl Shy, meant to be hilarious using cross-dressing actors. The all-male cast was drawn from the varsity football team. Deborah Hart directed it and cast me as Barbara Sanford the femme fatale. It was a role that I didn't want. It did nothing for my self-esteem as a male. And worse, the whole thing was partly my fault. I maneuvered myself into that silly play by suggesting the project.

My inability to excel in sports pushed me in the direction I didn't want to go. I had become a book-worm, a detested brain—no better than a sissy. My fantasy of supporting a college education with a football scholarship ended. For months I had nursed the fantasy that I could get a football scholarship and earn my way through college as several of my Mooseheart heroes and role models were doing.

Coach Powelson's assessment of my abilities revised my fantasies for the future. If I was too awkward to play basketball maybe I was smart enough to get an appointment to Annapolis or West Point where I could get a chance to play college football.

I applied to the Congressman from my district in Illinois for appointment to the Naval Academy. I received and studied the literature, which stressed the competitive basis of the appointments to Annapolis and to West Point. It was clear that I must take several courses in mathematics including a course in trigonometry.

Then a most unpleasant thing happened after I decided to stay at Mooseheart for another year. It started over a piece of gingerbread. Cliff and I both moved into Krebs after the seniors left. We really wanted to be in

Krebs because Mrs. Jones was the best cook and probably the best housemother in the privilege halls.

"First Extra" was a custom in most of the halls. When a guy called "First Extra" before a meal he was entitled to the first extra piece of fruit or dessert—if there was one. There often was when some guy missed a meal or Mrs. Jones made an extra dessert. Cliff and I got into an argument and then a fight over an extra piece of ginger-bread.

It was the most vicious fight I ever had. It happened on the hill where the water tower used to stand. The fight seemed to go on for almost a half an hour. Cliff was big-ger and stronger and pounded the devil out of me. I was able to keep poking my fist into his face and left him with a bloody nose and black eye. No one won. Neither of us would give in. Some of the kids watching must have called "Jiggers" because the fight eventually ended. We were not caught or reported or punished.

Mother was upset and cried. I was mad at Cliff and felt bad about the fight. I arranged to move from Krebs Hall back to Quaker Hall with Miss Pratt. She wasn't a good cook. But I didn't want to be around Cliff. We were never close, never good brothers—the way we should have been. Both of us got along with Clyde the way brothers should get along—even though he was just a little kid. It seems strange, but Earl was more of a brother to me and to Clifford, than Cliff and I were brothers to each other.

Cliff and I talked about the fight at one of the Mooseheart Homecomings not long before he died. He didn't remember that the fight was caused by an argu-ment over a piece of Mrs. Jones' gingerbread. And he was probably right; it was caused by my beloved Granny Pritt and Carrie Meyers. They used Blanche's sons as pawns in their struggle to influence and control "their daughter." They deprived Cliff and me of much of the affection and appreciation that we should have shared even though we lived a continent apart after we left Mooseheart.

Not long after the fight with Cliff, I received a letter from Christopher Droop, who knew our father when they worked for the Cherry River Boom and Lumber Company. He had visited us at Mooseheart a year or so before and I had guided him and his friend around the campus. He said he might be able to get me a job in Morgantown so I could work my way through West Virginia University and go on to the Law school there. I had mentioned an interest in studying law when I first met him at Mooseheart.

My mother had pushed me toward a professional career years earlier at Spruce Knob where she read poetry to me and urged me to become a Civil Engineer, a Doctor, or a Lawyer. Gradually it became apparent that at least I could please her, follow her wishes by becoming a Lawyer. I certainly was not interested in becoming a Physician. But a military career with an education tilted towards engineering and a chance to play college football continued to rule my fantasies.

Then suddenly those fantasies were wiped out by the Depression of the 1929-30's. I was in Miss Gale's office a few days after I received the letter from Mr. Droop. I don't know why I was in the office of the Supervisor of Girls. I suppose I was there to get a lecture about the way young gentlemen are expected to treat young ladies. But I do remember that it was there that I learned trigonometry would not be taught at Mooseheart next year because of the depression cutbacks. That ended my hopes to go to the Naval Academy or West Point to get a college education, and play football. I did not have enough mathematics to pass the examinations to either academy.

I wrote to Mr. Droop that I would be in Morgantown to enter the University before classes began in September. Mother records in her Diary and Daily Reminder that on a Friday August 25, 1933,"Ernie came in all excited. He has taken a notion to go back to W.Va."

Blanche records several attempts to get me to change my mind including a visit to Superintendent Roselle. She notes that he also thought I should remain at Mooseheart. On Tuesday August 29 she writes in the day book:

> Matron in Rochester. Was blue and nervous. Ernie has made up his mind to go back to W.Va. I had hoped he would change it. He and Cliff took the trunk over [to Quaker City] this forenoon.

On Wednesday [August 30, 1933] her entry reads:

> Letter from Earl. Matron in Dixie [hall] today. Was so blue and worried. Ernie left at 6:00 P.M. I gave him his daddy's watch and bought him a fob. The dear little kids in Dixie cried in sympathy with me.

She didn't record that she also gave me a copy of a book about sex called *Married Love,* by Marie Stopes, and a Manual complete with a diagram of the of female sexual organ and instructions on positions for sexual intercourse. She advised me:

> Read them. Always take care of yourself.
> I don't want you messing around with
> trashy girls. Remember there clean decent
> girls out there as interested in love as you
> are. And since you are Homer Rhodes'
> son I guess you'll find them. Treat them
> and treat yourself with respect.

My mother's was the only instruction about sex that
I was given at Mooseheart.

Suitcase in hand, I was waiting for the streetcar to
Aurora to catch the train for Morgantown, WV. Football
practice was over and Coach Seeglitz came by in his automobile and picked me up. It was lucky for me—I didn't
know where the railroad station was. I thought it was
at the end of the streetcar line. Anyway the Coach took
me directly to the train station. I did not tell him that I
didn't know where to find the railroad station.

However, he told me something that surprised me:

> I sent Cliff and Jim Heilman back to the
> second team today. And now that you're
> leaving, I will have to find someone else
> like Herb—with speed enough to get
> down under punts."

For the record: Herb was as slow as cold molasses.
But, Herbert Heilman rose from being the Governor of
the Junior Order of Moose lodge at Mooseheart to become the Supreme Governor of the International Moose
Lodge.

I didn't say anything to the coach about his news.

I no longer cared about him or his football team. But I wasn't displeased by it. I had other things on my mind.

On Tuesday [September 4, 1933] Blanche writes:

> "1st letter from Ernie
> Cook in W.C. [White Cottage]
> Had a lovely day. Was all excited, had a letter from Ernie.
>
> He has a job and is sitting on top of the world. . . ."

SCATTERING AND SETTLING

BLANCHE'S MOOSEHEART FAMILY began breaking apart and scattering when Earl graduated with the Class of 1932 to return to Carthage, Missouri, his original home. When Earl came back for graduation ceremonies in June 1933, mother would have five of her Mooseheart family together for the last time. Carolyn Guinn had not yet developed bonds of kinship with her brother's adopted family as Earl had, Those ties would not become evident until Carolyn and Bob Hanke embraced Clyde and began nurturing him after he graduated at Mooseheart and enrolled at Lawrence College in Appleton, Wisconsin in 1939.

Mother left Mooseheart after Clyde graduated in June 1939, but as long as Granny Pritt lived mother never settled into a comfortable home-life situation, except for her brief marriage to Howard Johnson. In her "Little Book" on September 6, 1939 she notes that:

> Life has been very hard for me lately. Miss Stover has butted into my life and made things almost unbearable. My work has been practically tripled. . . . I am going to see Dot who has moved to Alton, Ill. I

> will go to Mother Pritt. I am thinking of
> staying with her if I can see my way clear
> to our getting along. . . . poor Mother
> Pritt is so old and lonely.

She notes in later parts of the Little Book that she left Mooseheart on Sept. 20th, 1939. And She writes two entries in a row under a heading dated Cedar Grove, Nov. 5, 1940: " I came to Mother Pritt's on Oct.8, 1939". [In Mammoth, WV] "We moved to Cedar Grove on July 2nd, 1940."

Mother refused to move to be with, or near, any of her children as long as Granny Pritt needed her. Her insistence on caring for Granny Pritt made her endure the most lonely and miserable years of her life. Marcena would not move from the hovel on Williams Street. Blanche refused to leave the woman who had taken her in as an abandoned infant.

Even after Blanche married Howard Johnson, (Photo 13) Granny refused to budge from the house on Williams Street to live with them. Howard was a widower living a couple of blocks away in a relatively comfortable house with plenty of space. Granny Pritt had known and liked Howard most of her life, but stubbornly remained at Williams Street.

After two years and four months of marriage, Howard died of a brain tumor November 8, 1951 and Blanche moved back to Williams Street. When Marcena Pritt had to move into a Nursing home and died in 1964, mother moved to an apartment Dorothy and Earl prepared for her in their home on Elm Street in Centralia, Washing-

Photo 13 -- Blanche and Howard Johnson, marry July 1949

ton. She lived there briefly and died in a nursing facility in Centralia in 1966.

The last entry in mother's "Little Book" is dated July 30, 1952, less than a year after Howard died. It reads:

> I am 61 years old to-day. My life is so terribly empty. It is hard to get interested in anything. Things are so cruelly changed since three years ago to-day. Then I was started on a new life that looked so bright, and happy. Oh! why, did it have to end so soon? Why, oh! why?

Many women have suffered intolerably for their children and family, but I have known no one whose life was so beset with such misery, poor health, hard work, loneliness and misfortune as Blanche Rhodes. She sacrificed her life for her family, Homer Rhodes' children and Marcena Pritt. Abandoned by her birth-mother as a three month old baby, she would draw the short dirty end of the stick in almost every turn of her life.

Blanche's world centered upon Homer until he died. And then her attention and devotion shifted to her children--that is, HIS children. Of course she loved us, sacrificed for us, fought for us. But we were not the joyous whole of her being that was Homer Rhodes. We were just all that she had left of him.

Mother was a religious person and she was sustained by her faith through the loss of our father. Pictures of her before Homer Rhodes died often show her with a Christian cross around her neck. She told me that as Daddy was bleeding to death following an operation for stomach ulcers at the University of Maryland Hospital in Baltimore, she asked him to put his trust in Jesus. She said he grinned, and nodded that he understood. Then he died. Most of her died at that moment. And she would spend long periods for much of her life in the hospital from heart problems related apparently to rheumatic fever as a child and young woman. But she was steadfast in her religious faith when we entered Mooseheart.

She speaks to Homer in an entry in her Memory Book written in the Mooseheart Hospital and dated October 8, 1926, addressing him as:

Dear Pal of Mine:-

> Nineteen months ago to-day you were still with me and I, hoping against hope that you would be spared.

> But, God took you away from me, My Darling, and my life has been so sad and so lonely without you.

> For the heart that has truly loved, <u>never forgets</u>, but as truly loves on to the close. But, God has never forsaken me, my faith in him has been perfected, and though sometimes, it seemed as though life was not worth the living, still I have never doubted the "Divine Love." And, though I have been crushed, I have not rebelled . . .

> Our love was such a beautiful thing Dar-

ling, the trials of life meant nothing to us so long as we shared them together.

. . . . . .

Oh! Sweetheart, Oh I can not say and will not say that you are dead--you are just away. I have no remembrance of your last words, darling. But Oh! How I treasure the remembrance of your last beautiful smile and the slow inclination of your head.

You were ever an Optimist Bud, while I was a Pessimist. Your optimistic views and the wonderful love you bestowed upon me made life worth while.

Now, that you are gone Dear, nothing but my Religion has saved me from the blackness of Despair.

A week later, October 18, 1926, while she is still in the Mooseheart Hospital she writes:

Oh! Memory Book, how strange and far away it seems that October day, when I first held my second son in my arms. Oh, how much I have lived, loved and suffered since then...

. . . . .

Such terrible loss, that only the loving-kindness of my dear Heavenly Father has saved me from utter despair. I am trying to build a new life. But how strange it is to think of a life in which Homer has no part. But I must try, it is my Duty and I have no choice. God has given me many wonderful friends in my hours of need.

Blanche is still in the Mooseheart Hospital when she writes November 25, 1926:

My memory Book: This is the evening of Thanksgiving Day. My first Thanksgiving in Mooseheart. I feel with God's help. I can take up the burden of life again. Dr. Haslem brought me to church in his car. We had a beautiful Thanksgiving.

On March 8 1927 Mother writes of her work at the Campanile and in the Dentist's Office for Dr. Ghent. She is pleased with both jobs but uncertain about how long they will last. She recalls in this entry the anniversary of Homer's death (March 9, 1925):

It is two years to-morrow, [since Homer's death] Memory Book—two years—Oh! "Dear Heavenly Father," I thank Thee that they are over and done with. So much tragedy has entered my life—I have suffered so much Oh God , grant that I may never love any earthly being, not even his children —as I have loved him. Oh! God, please give me health and strength and interesting, absorbing work that leaves little time to brood, and Oh! dear Father, please keep me pure, pure and clean that Homer's children may never need to be ashamed of their mother. And, dear Father, help me to forget the mistakes of the past, and to press on to greater achievements today.

On April 12, 1927 Mother writes at the Campanile:

I am happier than I ever thought it possible to be. I am still in the Dental office and hope to be there permanently, as Mrs. Kissel isn't coming back and Dr. Ghent is pleased with my work. . . . Billy Sunday is preaching in Aurora now, I certainly want to hear him.

. . . .

The kiddies are growing up so fast. I hardly know whether I am glad or sorry.

This entry (on pages 24-25) in her Memory Book was last one she made for more than five years. She writes again on page 27 (dated Tuesday, August 2, 1932) of her health but there is no mention of her spiritual condition, her faith.

[It is} six years ago this evening since I arrived at Mooseheart. I hardly recognize

> myself to-day when I recall the pale frail
> little women of six years ago. My health
> is excellent today.

While I cannot speak for my brothers and sister, Mother influenced me in forming many of my ideas and in shaping my character. For instance, the only instruction I ever received about sex was from her advice, proscriptions, and recommended reading.

Perhaps one reward for her sacrifices was that the conduct of her children has generally conformed to society and its rules. Certainly, one of the most satisfactory experiences of her life was the informal adoption of Earl Guinn or his adoption of her. Earl became the fifth member of her family: a son to her, and a brother to her three sons by Homer Rhodes, many years before he became the husband to her one daughter. She was too worn out and ill to really enjoy the announcement that I had completed my PhD, other than say with a weak smile: "Now I have two Doctors in my family." Earl had completed his work as a Doctor in Optometry before I finished my BA in 1936.

Earl wrote to her regularly after he left Mooseheart and contributed financially to her care. Within a year after he graduated, Earl returned to see her—and doubtless Dorothy --at the graduation of the class of 1933. It would be the last time Mother would have all of her Mooseheart family in one place—but not under one roof.

When Granny Pritt had to be moved to a nursing home and died in 1964, Earl and Dorothy provided Blanche an apartment in their home on Elm Street in Centralia, WA. Like Granny, Blanche Rhodes was reluctant to leave that run down shack on Williams Street in Cedar Grove. Ironically, Earl had to stay in Centralia to take care of her grandchildren when she was buried beside Homer Rhodes on a dark miserable day in November 1966.

Thus Blanche and Homer Rhodes were reunited in adjoining graves on a hill at the outskirts of Richwood, West Virginia—-forty-one years after he died.

Clifford left before his class graduated in 1934. Our mother marched across the stage during the graduation ceremonies for the Class of 1934, accepting two diplomas for her absent sons. I was in summer classes, a sophomore at West Virginia University. And Cliff was at work on a roof somewhere in Los Angeles. I don't know what happened to my Mooseheart diploma. I've never seen it but it should have been dated 1933. I doubt if Cliff ever saw his. Mother made a practice of appropriating such items as diplomas for safekeeping.

Cliff settled down more quickly than others in our Mooseheart family. He enrolled in classes offered by the American Institute of Banking In 1935. After finishing he began working at a branch office of a bank in Los Angeles. He writes in 1936 that he was tending customers at a window of his own--with his own name plate.

Clifford married Patricia Ireland on October 12, 1941; Patricia Lorraine Joy Rhodes. Cliff and Patricia Ireland's only child was born on March 30, 1943. By 1950 Clifford had settled permanently into Banking. His career spanned nearly half a century with the Santa Barbara Savings and Loan which later became Santa Barbara Savings—retiring from that firm in 1998. Clifford's wife

Patricia Ireland Rhodes died from esophageal cancer in January 1974.

Dorothy graduated in 1935 and joined Mother and Clifford in Los Angeles. Dorothy finished an accelerated course at Marinello Cosmetology School and went to work for the May Company in downtown Los Angeles. Earl Guinn graduated from the Southern School of Optometry in Memphis, TN, in January 1936. With the kindly help of the founder of the School, Earl was able to work his way through the program and pay most of his expenses. Firing the furnace was one of his many odd jobs for the college.

After graduating, Dr. Guinn and Dorothy married in Los Angeles on her 19th birthday June 4, 1937. (Photo 14) Their first child Barbara Joanne was born in Illinois on June 4, 1940. A second daughter, Eileen Cheryl, was born on May 10, 1945, also in Illinois. Eileen was four months old and Barbara was five years old when their parents left Illinois and took a train for the West Coast, where they moved about frequently before finally settling in Centralia, Washington.

Their son Jon Guinn was born in Centralia on August 10, 1953, and the family rooted itself firmly in the community where Dr. Guinn's professional career flourished until he retired in 1982. Earl would become the most financially successful of Blanche Rhodes' Mooseheart family.

Throughout his professional life Earl Guinn was an active civic leader in the State of Washington and the several communities in which he resided, particularly in Centralia. Dr. Guinn was listed in "Who's Who in the West," in 1971 and 1972.

*Photo 14—Dr. Earl W. Guinn and Dorothy
Rhodes, marry June 4, 1937*

Clyde, the youngest of Blanche's family, graduated and left Mooseheart in 1939. He entered Lawrence College the same year, financing his studies in part with a football scholarship. Mother and I were the only members of his family to see Clyde play football for Mooseheart. Such a shame! Babe Rhodes was barely five feet six inches tall and at times weighed as much as 147 pounds. Even so, he was arguably Mooseheart's greatest quarterback before the legendary Harry Childress in 1950. Clyde's coach Johnny Williams, himself once a Mooseheart quarterback, is reported to have said so. Clyde a "triple threat player" could pass and kick a football as well as carry it in high school and later at Lawrence College.

Clyde told me that playing college football was extremely demanding. He said that he would often take such a beating in the Saturday games that he could not dress until Wednesday to practice for the next game. He admitted that he was relieved in his senior year when he was "scabbed out" (his position taken over, in part) by a 200 pound freshman. Clyde graduated from Lawrence in 1943.

After training in Miami, Clyde was commissioned an Ensign in the United States Naval Reserve and assigned to a Submarine Chaser in the Pacific. Our paths crossed twice during WW II. On one occasion I spent a day with him on a sub-chaser in Hawaii. And we were together briefly after my ship returned to Hawaii following the invasion of Okinawa. Before the war ended he was given command of a submarine chaser and officiated as its Captain when our government turned it over to Russia, to encourage that country to enter the war against Japan.

He was separated from the Navy in October 1946, and in March 1949 he moved to Seattle where he was employed by The Atlantic Insurance Company. For the rest of his life, he lived in Seattle except for a short time in Vancouver, B.C, Clyde and Alvina (Hill) Van Slyke married on September 1, 1951 and settled into a home of their own. Although they had no children, Alvina had an-eight-year old daughter, Penelope. Except for blood, Penelope Rhodes became in every respect Clyde's only child.

Clyde died in December 1980, following a long painful struggle with cancer. I believe he was the brightest and most loveable of Blanche's sons even though she sometimes joked that Clyde could be "the most contrary of her kids."

RETURNING AND REMEMBERING

CLYDE NEVER RETURNED Mooseheart, not even when he was inducted into the Athletic Hall of Fame. And I did not settle into my profession or a real home until after Earl's sister, Carolyn, became the Matriarch of the Rhodes/Guinn/Hanke Tribe from Mooseheart. (Photo 15)

Carolyn Guinn married her Mooseheart sweetheart Robert Hanke in 1934. Bob was creating a reputation as the Swimming Coach at Wauwatosa High School in Wisconsin. By 1938, his teams were in a nine-year stretch of 102 victories in dual meets, three state swimming titles, and the development of swimming stars including ten All-American Interscholastic swimmers and divers. Coaches throughout Wisconsin would listen to him about high school athletes with talent.

Robert noticed that Clyde was an excellent football player and mentioned him to a coach at Lawrence College, in Appleton, Wisconsin. When Clyde enrolled with a Football Scholarship at Lawrence in 1939, Bob and Carolyn's home in Wauwatosa, Wisconsin, became Clyde's only home. Clyde found another big sister and a little mother in the Matriarch of the Tribe from Mooseheart. And Bob became another big brother and parental role model.

*Photo 15 — Matriarchs of Rhodes/ Guinn/
Hanke tribe from Mooseheart ca, 1960*

Robert Hanke became the Superintendent of Mooseheart, in 1974. (Photo 16) Earl and Dorothy began spending their visits to the campus in Father Laffey's former house which had become the Superintendent's residence.

When Robert Hanke, Carolyn Guinn Hanke, and Clyde Rhodes all died in 1980, my sister Dorothy became the Matriarch of the Tribe. Homecoming at Mooseheart became the occasion for an annual reunion for the four survivors: Earl Guinn, Dorothy Rhodes Guinn, Clifford Rhodes and me. My sons Stanley and Lloyd usually accompanied me and sometimes Earl and Dorothy's daughter Eileen joined us.

Charlotte Rapp's newsletter, *The Rapp Sheet,* helped us to keep in touch. We usually met in the Lobby of the Howard Johnson Motel in North Aurora the second week-end of each October. Those meetings with class-

Photo 16—Superintendent of Mooseheart Robert Hanke and Carolyn Guinn Hanke

mates and other former Mooseheart students continued for the next two decades into the new Century.

After Clifford died in 2002 and Earl died in 2006, Homecoming became a joyless occasion. Dorothy and I stopped attending. We simply have no one left share hours of talk about the old days. With family deaths and the loss of many other Mooseheart classmates, we had no one who would comment, as if the ideas were theirs: "I owe almost everything I am or will be to Mooseheart" and claim that "Mooseheart is the only real home I ever had that is still standing."

Those thoughts of gratitude have wavered into and out of in my consciousness since I graduated from college and became a member of the General Assembly Lodge at Mooseheart. I certainly felt them clearly on a Sunday afternoon in June 1942 when I returned to Mooseheart for the first time after leaving nine years before in August l933. (Photo 17)

Much had happened during those nine years. I had graduated in two years and nine months (June 1936) with a BA degree from West Virginia University. After selling subscriptions a few months for the *Northern Virginia Daily*, in Strasburg, VA, I was hired as reporter-in-charge of their Winchester branch office. I married Helen Mason in September 1939. The newspaper fired me as "temperamentally unsuited" about a month later. My father-in-law found me a job in a woolen mill where he was employed. Roland Q. Nicholson, with whom I had worked on the *Northern Virginia Daily* rescued me from the woolen mill in October 1941. Nick got me a job reporting for the *Alexandria Gazette* before he moved to the *Washington Star.*

Photo 17 -- Reunion of Blanche's Mooseheart Family in 1942. Blanche, baby Barbara Guinn, Dorothy Rhodes Guinn. Second Row, l to r, Clyde Rhodes, Ensign Ernest Rhodes, Dr. Earl Guinn.

About 9 a.m. on April 9, 1942 I walked out of the *Gazette* onto King Street with a manila envelope in my hand and boarded a street car. Soon the car crossed the Potomac River, at that time the largest body of water I had ever seen. It stopped in front of a Navy building in Washington, D.C. I debarked, started into the building, and a Marine guard stopped me, asked to see the manila envelope, opened it and glanced over the contents. Then he snapped to attention and saluted, "Ensign Rhodes, you are on duty, Sir!" For most of the next eighteen hours I pasted strips of paper containing secret information onto half-sheets of paper.

The Navy needed commissioned officers to handle secret messages. In less than a month later, I found myself bored with handling secret messages and applied for sea duty: "Application Granted." I was ordered to Tower

Hall in Chicago for thirty days instruction in seamanship, navigation, and ordinance. I needed it.

I had no idea of going out to Mooseheart for a visit, although it was only thirty-six miles southwest of Chicago. I had no way to get there until I met a former Mooseheart graduate then living in Chicago. Vance Davis a former classmate of my sister, Dorothy, drove me out to Mooseheart on Saturday in time for the annual graduation ceremonies.

Vance had promised to get me back to Tower Hall before lights out on Sunday. He was late. As I waited for him at the pylon (Photo 18) at the entrance to Mooseheart, I did not consider the possibility that in a few days I would say goodbye to Vance forever. He would join the Air Force and be killed in a plane crash in Texas shortly before the war ended.

I was preoccupied on that memory- filled afternoon with excitement about my orders to a ship in the Pacific the *USS President Jackson*, with apprehensions that my abilities and courage might not measure up. Mixed with these feelings was the the realization of how much I had missed Mooseheart, For the first time I sensed that I loved this place where I spent half of my childhood. I wondered when I would again pass this pylon; I wondered when I would return to the only place then standing where I had been fed, clothed, sheltered, trained, educated, and valued. It occurred to me that I had no home still standing but Mooseheart.

My home with Homer and Blanche Rhodes at Spruce Knob was by June 1942 nothing but a disappearing scar on the side of Fork Mountain in West Virginia. Certainly the house of mean-spirited Carrie Thomas and

Photo 18—Entrance to Mooseheart

George Meyers was never my home, and could never be my home. Never! The apartment I furnished with Helen Mason across from the Palace Theater in Winchester, Virginia, was a temporary residence to which I rode a bus from Alexandria for weekend visits.

That Sunday in June 1942 as I waited for Vance at the entrance to Mooseheart, I did not suspect that for years following the war, I would become an academic nomad, moving from college to college--finishing my education and accepting temporary teaching opportunities—until I eventually settled in Norfolk where I remarried in 1969.

While waiting for Vance, I felt (probably for the first time) that I owed the Moose Fraternity and Mooseheart, a debt. I now believe that it is an enormous debt—a debt that I can never repay, one that I can only acknowledge.

The Fraternity has never asked me for one penny in return for its benevolence. A most unusual benevolence: with one exception, to my knowledge no Mooseheart child was ever struck by an adult. The offending night watchman did not see the sun rise as an employee of Mooseheart. Perhaps just as remarkable, Ernest Rhodes was never verbally abused by an adult (although he was often lectured with exasperation). Every child was given all that was needed to stress its individuality, preserve its sense of worth, and prepare it for almost anything it was capable of doing, or becoming.

At the Homecoming in 1990 I would finally reach a satisfactory conclusion about my debt to Mooseheart. Also I would have been settled for over thirty years in a home—a place from which I could be removed only by fire or flood or else carried out feet first. I was removed by flood in 2003 and the house is no longer standing.

Earl Guinn, already a member of the Athletic Hall of Fame, was honored by the Alumni Association in 1989 as a Distinguished Mooseheart Alumnus. Coach Johnny Williams, Russell Raycroft and I were named for the same award to be made at Homecoming in 1990.

In my acceptance talk I paid tribute to the three most important women in my life: Blanche Rhodes, Dorothy Rhodes Guinn, and Dr. Carolyn Hodgson Rhodes, my colleague, my best friend and my wife. The occasion went smoothly. My Mooseheart sweetheart Meg Koski was present and came by to congratulate me and whisper: "I always knew that you were special," I accepted her compliment --with the knowledge that it was "stretched." The most important thing that happened that night was meeting a former student. He became a successful lawyer and an early recipient of the Distinguished Alumni Award.

He congratulated me with the observation: "It's about time they pay attention to teachers." In the small talk that followed the question came up-- as it often does with former Mooseheart kids: What do we owe the Moose?

I believe I have paraphrased the Lawyer correctly:

I know I can never repay my debt to the Moose in cash. Of course, the Fraternity will probably accept money, if we want to try to repay for what we have been given. I doubt if any Mooseheart Kid has ever tried. But the Fraternity expects much more from us than cash.

Have you ever noticed the words carved on the low wall at the right of the pylon at the entrance to Mooseheart? Those words say in effect: For all we have been given the Moose expect service from us to society-- the very best we can give.

ENTER TO LEARN   LEAVE TO SERVE

BLANCHE PORTER RHODES

GRANNY PRITT TOLD me that she saw the woman who abandoned our mother get off the train from Charleston. The woman was well dressed and good looking. She had a three-month old baby in her arms--a baby girl who later became our mother. The woman was alone with the baby, and no one met them. Apparently she found her way to the only hotel in Handley, West Virginia, a small town along the railroad that ran east from Charleston toward Clifton Forge, Virginia. It happened, more than likely, late in October or early in November 1891.

Granny Pritt told my sister Dorothy that someone asked her to take care of the abandoned child until its mother could be located. Granny says she found the baby covered with sores in one of the hotel rooms in Handley with all of the windows flung wide open "It 'pears to me like someone left the baby there to freeze to death," Granny speculates telling the story years later. I suspect that the woman who abandoned the infant left the windows open purposely to cause the baby girl's discovery--as it did. In any case, Granny Pritt was called to care for the baby. She took it into her home. After the mother could not be found, she decided to keep it. She named the baby Blanche Anis Porter and reared her to adult life.

Granny Pritt was born in 1868, in Liberty, Putnam County, West Virginia. When she first saw the infant that would become our mother, Granny was married to William Lucian "Buck" Porter, a blacksmith living in Handley. Buck died in 1906 and Granny married Charles Pritt. She had no children by either husband. Our mother would become her only child. Marcena Pritt was the only mother Blanche knew until after she married. (Photo 19)

Granny Pritt was an easy-going, poorly educated, snuff dipping, kind-hearted soul. I worshiped her. And she clearly felt as if I put the sun up for her most mornings. No question but I was the favorite of her daughter's children.

I regret that I never pressed Granny Pritt or my mother for details about Blanche's birth and abandonment. But in my mother's day an unmarried female who gave birth to a child was considered sinful, or at best a weak, stupid, common slut. And the child of such a woman was viewed as "trash." The subject of illegitimacy was taboo, especially in families where it had occurred. Children were shielded from most aspects of pregnancy, sanctioned or otherwise. Consequently, many of the simplest details that Dorothy and I have tried to piece together about our mother's birth and abandonment are missing.

While Blanche was troubled by the stigma attached to her illegitimacy, I don't believe she was ever bitter toward her birth mother. Blanche was living in Los Angeles when she writes in her Memory Book on July 30, 1936:

> My 45th birth-day. I was born in a small
> hotel in a little village in W.Va. Poor

Photo 19-- Granny Pritt

little atom of humanity. Unwelcome, unwanted. Poor unhappy young mother. Betrayed by the one she trusted most.

> Small wonder that she grew desperate and
> deserted her baby when it was 3 months
> old. No human could have started life
> under more tragic circumstances. B.R."

Neither my mother nor Granny Pritt told me much about my mother's childhood. She apparently finished grade school. Throughout her life she regularly wrote informative letters, clear and easy to read. Her handwriting was even, smooth, beautiful, I thought. We know she was once sick, very sick from rheumatic fever, when she was about thirteen years old. It left her subject to heart problems until she died in 1968. Rheumatic fever struck her again shortly before she married our father. He visited Blanche at the Sheltering Arms Catholic Hospital near Charleston. I saw a postcard he wrote to Granny Pritt reporting that Blanche was improving. But apparently it has been lost.

Blanche Porter married Homer Rhodes in Ward, West Virginia, on June 1, 1910. (Photo 20) For years she kept a small tin box containing ten new Lincoln head coins dated 1910 that my father gave her on their wedding day. I believe it was the only wedding gift exchanged--other than two priceless gifts: themselves.

Our parents' attempt to start a family began sadly. A daughter Shirley Lillian died at birth on August 16, 1913. Dorothy told me that mother was suffering from her heart problem when the child was born, and that for much of her life mother carried a sense of guilt about the child's death. Such a feeling of guilt for surviving is related, I suppose, to caring—to loving too much. Most of us have probably heard a parent or an older person

Photo 20– Blanche Porter and Homer Rhodes marry, 6-1-1910

lament—"why couldn't it have been me—they had a whole life ahead of them."

I was Blanche's second child. I was born on April 17, 1915, in a log cabin formerly used as a schoolhouse and located at the head of Rich Hollow near Mammoth, in Kanawha County. My father was working in a coal mine near Mammoth at the time.

After I graduated from West Virginia University, in 1936, mother took me to the site of the log cabin where I was born. It had burned down or been taken down shortly before. There she retold the story about the time she left me on a blanket at the cabin door and returned to find me crawling toward a copperhead snake that moved to be near me. She said she beat it to death with a broom. And again she jested that she picked out the log cabin as the place for my birth so I could have a head start on most other candidates when I was old enough to be president.

Dorothy says that she, Clifford and I were all born in the same log cabin. Clifford Lawrence was born on October 15, 1916, and Dorothy Virginia was born there on June 4, 1918. I would feel perfectly confident in nominating her to be the first woman President--and so would the rest of her family. (Photo 21)

Cliff and I were destined to become sibling rivals. He grew to be taller and stronger than me, and I perceived him to be much more handsome. From the start I resented him feeding at "my bosom" on my mother's body. Mother reports that she once offered me the breast on the other side from which Cliff was feeding. I took it. When father came into the room and saw his sons hanging onto his wife's nipples, mother quotes him as saying: "For Christ's sake, Blanche, stop that! You look like an old sow."

Mother and father moved out of the log cabin in

Photo 21—Blanche and Her Log Cabin Babies

Rich Hollow after Dorothy was born in June of 1918 and before we moved to Spruce Knob in 1919. But I do not remember any of the houses we lived in until we moved to Spruce Knob. Our home there must have been much better than the log cabin in which the first three of mother's children were born. We were certainly better off economically for Homer Rhodes had risen from a miner swinging a pick and loading coal to foreman and then to superintendent of a coal mine.

Mother worked hard and was often ill at Spruce Knob. Even so her time there before our father became ill was probably the happiest time of her life.

Carrie Thomas, Blanche's birth mother, reappeared and inserted herself into her abandoned child's life shortly after Blanche and Homer married. Blanche's acceptance of Carrie Thomas Meyers as Mother Carrie was a mixed blessing. And surely the antagonism between her two mothers distressed her. But I believe she is talking about those years at Spruce Knob when she writes in her Memory Book from the Mooseheart Hospital on October 15, 1926 that: . ." I have had many days of glorious happiness and many golden dreams, some of which came true, followed alas! By heart-breaking tragedy and loss."

Mother's child-bearing ended December 31, 1920 in her comfortable home at Spruce Knob. The baby who quickly became everyone's pet was named Clyde Alan Rhodes. Clyde grew to become the most independent and self-sufficient of Blanche's children.

After Clyde was born and before Daddy became ill, I began to be aware of the partnership that existed between my parents. I picked up information about the mine and my father's work from overhearing their talk as he shared his problems with her. They discussed everything from Charlie Pritt's drinking to the opening of a new seam of coal inside the mine.

Mother was spirited in her role and conscious of her

obligations as the partner of the superintendent of the mine. She was also a Baptist, and concerned with her Christian duty. In an illuminating flash she felt called to start Sunday Afternoon Worship Service in the schoolhouse. Daddy humored her, even though he often indicated that his faith did not concern itself with religion or with an afterlife--at least in his present suit of flesh and skin. He would tease her: "When I die, just throw me out to the pigs."

Nevertheless, Daddy found twenty-four hymn books for her. When Sunday afternoon came Blanche shepherded her husband and their children into the schoolhouse. The hillside wasn't teeming with devout Protestants. Blanche's family and a few of her closer friends had the place to themselves where they sang When the Roll is Called Up Yonder. And that one meeting ended, as far as I can recall, the not very Great Awakening at Spruce Knob in the early 1920's.

Their way of dealing with each other and me can be shown by their treatment of my question about the meaning of the word fuck, which I heard while playing out on the motor car track.

"Don't you worry about it son, we'll tell you all you need to know before you are old enough to enjoy it," Daddy quipped.

"Homer," she warned icily in tone that said that's enough of that.

Turning to me, she explained:

> It's something that concerns older people
> and animals. You come to me and we
> will talk about it, whenever you want--

whenever we have the time. Right now we don't have the time! You're on your way to the powder house. Bring me the empty boxes. We need kindling to start the stove. We've got to fix supper for your Daddy and the rest of us. Now git!

I later found the time to ask mother about "fuck" and she calmly told me about the birds and bees. She touched on the difference between male and female. She noted some of similarities of dogs, cats, and cows to other animals, even whales. She called them mammals. She revealed that babies were made in the stomachs of females and were fed milk by their mothers.

Not long after, Daddy took me with him when he led our cow Daisy out to the flat near the mule barn. Emory brought out Caesar, his big bull, with a rope tied to a ring in its nose. While Daddy held Daisy, Emory let Caesar get on top of Daisy's back. Then Emory pulled her tail away from where she peed. Caesar started jumping around and Emory took the bull's long skinny pink looking thing and put it in Daisy.

"It's to make her have a baby cow, and when she starts making milk to feed it, we'll get our share," Daddy told me. He didn't try to explain what the word "fuck" had to do with the episode between Caesar and Daisy. And he didn't suggest to me that it was fun for either the bull or the cow.

One of my jobs was to bring Daisy home each night to be fed and milked. I would wander along the motor track before the sun started down listening for the bell

Daisy wore around her neck. Then I would go into the woods, pull a switch from a bush, and drive her home.

When the proper time passed after the episode between Caesar and Daisy, I found a calf with Daisy and drove the mother and baby home.

"Its mine, I found it," I claimed. Daddy looked at mother. He always looked at mother, when I asked for something. She answered him as she often did: without moving her head or saying a word or even blinking an eye.

"All right," he said, "it's yours, but you have to bring it home every night and pen it up."

"Only until it's weaned," mother stipulated, "only until its horns start growing."

"You must bring it home, feed it, and pen it up under the house, every night, or else that calf is your mother's and mine. Every night," he repeated.

I asked, "Every night until it gets hookers? It's a baby. It don't have hookers!"

"It'll get them," Daddy assured me.

"And then we'll have to sell it," mother announced.

"It's mine," I argued, "I don't ever want to sell it."

"We'll see," Daddy temporized.

"We can't keep it around when it grows up and gets horns. It can hurt you; it's a bull and they can get mean, like Caesar," mother explained.

I kept my part of the bargain about feeding the calf. Every evening before dark, I would go out find Daisy and drive her and the calf home. Daisy would head for the pen under our house where she could be fed and milked. Often the calf had to be coaxed into the pen by a pan of feed soaked in milk. He would run excitedly after me as

I dragged the pan downhill along the ground. Finally the calf got so big and fast that he could outrun me. Once he trapped me against the side of the pen before I could get out of the way. Since the little bull's horns were beginning to break out on the side of his head, mother ruled: "That's it! We're selling it."

I puckered and sniffled. But really, I didn't feel bad. I had scraped some of the skin off my side and belly when the calf trapped me against the pen. And actually, I was afraid of it, and was relieved when Daddy said: "The calf's yours; so we'll put the money we get for it in your account in the Citizen's Bank."

They did.

The bank failed during the depression of 1929.

Daddy and mother regularly found jobs for me, and rewarded me for doing them: sometimes with money, but more often with praise. By the time my father became sick with stomach ulcers, mother had made me into her "Little Man" and said she hoped that I would grow up to become a good man--like my father.

After we ate supper she would lift me up on an empty wooden powder crate in front of the rinse pan, pull the washed dishes from the scalding water, and I would dry them--just like daddy did when he was feeling good. I was not expected or required to do it, of course not; it was a privilege that I was allowed because I was my father's oldest son.

Daddy's illness and mother's needs drew mother and me closer together. Sometimes she would have one of her fainting spells. Her nose would twitch like a rabbit's nose, and she would sit down or lie down on the ground. I would run, find her spirits of ammonia, and hold the bottle to her nose until she could help herself up. Then I would get her a glass of water and her little red pills and help her to a place where she could rest and regain her strength.

Blanche Rhodes valued education. As long as she lived, she read books and kept them about the house-- especially self-improvement books. One that I remember dealt with teaching yourself English--how to read and write. Some of her earliest gifts to me were *A Child's Garden Verses* by Robert Louis Stevenson and a King James Version of the Bible. She loved the long poem "Snow Bound" by Whittier, and read me poems from a book she owned by Paul Lawrence Dunbar. She encouraged me to memorize poems like "The Village Blacksmith" some of which I can still repeat--probably inaccurately:

> Under the spreading chestnut tree
> The village smithy stands. . . .

Or one that I rather liked was:

> There was a little girl
> Who had a little curl
> Right in the middle of her forehead
> And when she was good
> She was very very good
> But when she was bad she was horrid.

And then, there was a poem about a banner with a strange device "Excelsior."

Long before I was nine years old mother had given me my choice of becoming a doctor, a lawyer, or a civil engineer.

Once I was made aware of the fierceness with which Mother would defend anything she valued. In this case it was nothing more important than our Rhode Island Red rooster. He was important to us because we discovered him late one night back under our house covering a brood of chicks a hen had deserted because they were old enough to take care of themselves.

One morning a ferret (or weasel), a small light tan animal, fastened itself to Old Red's neck and the chicken ran into the yard where I was breaking up some empty powder boxes for kindling. I pinned the ferret to the ground with a piece of the broken wooden box. I called to mother. She ran out of the house, and saw what was happening. She grabbed a piece of wood and beat the ferret until it released Old Red. Then she continued to beat the animal with an awesome ferocity until it was just skin and guts.

We ate well at Spruce Knob: We had milk and buttermilk from Daisy and eggs from our chickens. We grew onions and I believe our own potatoes and once Daddy bought a sack of rutabagas; we had bacon, oatmeal, and cream of wheat. Mother made biscuits and fried pies filled with apples or peaches. She made ordinary pies, and cakes. She prepared gelatin with canned peaches in it, or sliced bananas and English walnuts. We had apples, applesauce, and fried green apples.

Daddy once brought home a domestic turkey that he won in a shooting match. He had pictures made of himself holding the dead turkey in one hand and his shotgun in the other trying to tease us into believing it was the biggest wild turkey that he had ever bagged. (Photo 22) We would occasionally enjoy a grouse from the woods or

Photo 22--The Meat Hunter.

trout from the Cherry or the Cranberry Rivers. We ate fried chicken and chicken and dumplings frequently, and at times meat--when people at the mines butchered. And I remember that daddy once bought home some cheese--

cottage cheese that had been salted, covered with a cloth, and left out in the weather to cure.

Mother used a lot of coffee (for the grown-ups), lard and soda. Arm and Hammer soda came in the same kind of yellow box it still does. But at that time, the box often included a card in it that pictured a common bird. The first bird I ever identified was a Junco. I recognized it by its picture on a card from a box of Arm and Hammer soda. Years later, I became a birdwatcher. The slate colored junco is the first bird on my life list.

Several times we made ice cream. In freezing weather, Cliff and I would take our sled to the reservoir near the main entry to the mine and haul ice home. Mother made vanilla custard and put it in an empty lard bucket. Daddy would put the lard bucket in a wash tub, pile ice around it and sprinkle on salt to make the ice melt. Cliff and I would try to relieve him of the task of turning the bucket around and around, while Dorothy and Clyde carried ice, until the custard mix froze into ice cream.

But we never made ice cream together after March 9, 1925, the day that Daddy died.

CHARLES HOMER RHODES

HOMER WAS NEVER addressed as Charles. We believe he lived exactly 35 years and 6 months. But we have no records of the birth or the adoption of our father. Homer Rhodes's vital statistics, like the records of his wife, and three of their children, may have been destroyed when the Kanawha county courthouse in Charleston burned—presumably before 1917.

Mother told me that my father was born on September 9, 1889, in Charleston, West Virginia. She said his mother, Susan Payne, gave Homer to Isaac P. Rhodes and his wife Annie when Susan married. Susan also had a son named Isaac Rhodes whom she surrendered to Bennett Rhodes, the brother of Isaac P. Rhodes.

My uncle Isaac visited me shortly before he died early in the 1960's. Uncle Isaac was of the opinion that Bennett Rhodes was his natural father and that Isaac P. Rhodes was Charles Homer's natural father.

This I know: my father strongly resembled the carpenter, Isaac Rhodes, the man I called Grandpa Rhodes. We visited him in 1926 on our way to Mooseheart after my father's death. He cuddled me against his side as we sat in a swing on the front porch of the house on Williams Street in Cedar Grove. He built the house on one of three adjoining lots—two of which had been originally purchased by Carrie Thomas and deeded to my mother.

Grandpa Rhodes pointed to a rocky outcrop on the side of the mountain that could be seen from the front porch. He told me: "We can find rattlesnakes up there sunning themselves on almost any clear day." That was the second time I had seen him; and it was the last time. I felt then and I feel now that I belonged to him--that he belonged to me.

My mother told me that our father finished the third grade and went to work as a Trapper Boy when he was twelve, opening and closing doors to keep air flowing in the mine.

Homer was hired in 1919 as foreman of the Elk Lick Coal Mines, above North Bend on the Cherry, about ten miles from Richwood. Unquestionably the job was obtained through the influence of Blanche's birth mother Carrie Thomas Meyers and "Daddy George" Meyers her husband. Daddy George was a person of some status in the Cherry River Boom and Lumber Company which operated the mines to power the lumber company's saw mill in Richwood. Carrie had reappeared in Blanche's life after mother married our father. She began immediately exercising control over our family's life through a number of obvious schemes.

As long as I remember, our father was superintendent of the mines and Lon Spencer was his foreman. They often studied rolls of blueprints, spreading them out on a shelf-like table that ran along one side of the small building which served as Homer's office. (Photo 23) It was very close to our house. I remember them talking about cars, empties, entries, openings, tonnage, the new battery

*Photo 23-- Homer at His Office Door with
Kids—Ernie, Cliff, Joe Postek, Dorothy.*

motor (that seldom worked) the mules, the monitors, the
tipple and needing to put a brattice some place in the
mine.

Usually I could understand from context what they
were talking about except for the term "brattice" (like
radish) and "bratticing." Not until I was in college did I
discover in a dictionary that a brattice was a partition or
wall of planks or cloth used to direct the flow of air in a
mine and that bratticing meant installing such a partition.

I remember my father for his sense of humor, his devotion to our mother, his affection for their children, and
his sheer enjoyment of being alive. "Don't you fool with
me," he would warn us "I'm rough and I'm tough. I can
slide down two miles of barbed wire fence--with a wildcat
under each arm--just for exercise before breakfast." If he

saw something pretty, he would often describe it as "cute as a speckled pup under a red wagon."

He enjoyed posing as a cook. He once took Clifford and me on an overnight trout fishing trip on the Cranberry River. For supper he fried potatoes and onions to go with the trout we caught. He urged us to tell mother he was a better cook than she.

We did.

She smiled.

One Christmas mother made an egg custard dish, dressed it up with sliced bananas and English walnuts, and set it out overnight to cool. He brought it to us in the morning and told us that it was Paris Pudding. He insisted he made it. Then he urged us to taste it and ask our mother why she couldn't make a dessert as good as his Paris Pudding. Of course, we did. She just smiled.

They worshiped each other. They shared private jokes and secrets. Even though she was sometimes ill, mother was spirited and enjoyed teasing her husband as well as accepting his teasing.

She might, for instance, allude to an incident supposedly occurring before they married: "Well, I never left my shoes under anyone's bed."

"I can explain that" he protested.

"Do!" She would urge him, "I'd like to see what you can add to the story."

"She was sick" he would begin.

"By she, you mean Eva Coster, George's wife? And the poor thing was sick--that's why she was in bed?"

"She wasn't in bed; she was sitting on the side of the bed".

"Without a stitch on her back?"

"She had on a night gown and some kind of a coat over it."

"So she was cold?"

"I don't know, but she had a cold."

"And you were in her bedroom to keep her warm?"

"I went to her house to take her some sassafras root to make tea for her cold."

"Last time you said you took her one of your old flannel shirts and some goose grease because she had a cold on her chest."

"I took her all of those things--she had a cold."

"And you left your shoes under her bed so she could keep her feet warm?"

"She put them under the bed to keep George from seeing them. You know how jealous George Coster is."

"In the first place, why'd you take your shoes off?"

"I didn't take them off!

"No?"

"No! Like I always told you, George is jealous. When I heard him come in I just jumped right out of my shoes--on my way through the window. You know how quick I am."

"How did you get your shoes back?"

"I don't want to talk about that. Everybody knows how that happened."

"But I like to hear your side of the story," she urged him.

"Well, George come up to me in the company store carrying my shoes. He dropped them in front of me and said 'I believe these are yours.'"

"And what did you say, dear?"

"Didn't say anything, he's a mean son-of-a-bitch; I'm lucky he didn't shoot me."

Daddy fished for trout and hunted, both to put food on the table and to enjoy the sports. Behind the table where we ate were two calendars: one pictured a version of the peaceable kingdom showing wild animals--several natural enemies driven together on a log spanning a stream, while a forest fire raged about them. The other pictured a dark haired fisherman, who resembled our father, excitedly indicating the size of a large fish with his hands held at least two feet apart. The title under the picture was "The One That Got Away."

As long as he lived, Daddy would repeat the story of the large rainbow trout that brother Cliff hooked and brought to the surface of the Cranberry River. He insisted: "the biggest trout in West Virginia got away from the smallest fisherman in the State."

Homer didn't pay much attention to the dates the hunting season opened or closed. But he always bought hunting and fishing licenses. I recall he once sneaked a grouse home in the back of his hunting coat and told mother to be sure to bury the feathers--it was the day before the season opened.

Late one evening, he took Clifford and me out in the woods beyond the mule barn. He was after a grouse that had been drumming to attract a mate. He left us in a clear spot: "Don't make any noise, and don't move from here. I don't want to shoot you." As it grew darker, we could hear the bird drumming, and all kinds of noises in the woods around us.

We began a series of low whines: "Daddy! Daddy, where are you?"

The drumming stopped, and Daddy thrashed through the underbrush to us. "You damned babies! I could have gotten both of them with one shot!" Never been much of a question about my father as a sportsman. He was a meat hunter. (Photo 22).

Daddy was a generous provider of good things for his family and himself. And as Superintendent of the mine, he made good money in comparison with the average miner. He bought a Howard railroad watch which cost him about $90. That was a lot of money at that time.

My earliest memory at Spruce Knob was of being in bed and my father standing beside the bed swinging that watch before me to get my attention. It was during the winter of 1919. And I was recovering from the flu which killed many people that year. Mother was standing beside him and they were encouraging me to "Wake up! Honey, wake up!"

He purchased the watch from a traveling peddler who made frequent trips to Spruce Knob. I remember the peddler selling Daddy a Jew's harp, a harmonica, and an inexpensive green backed issue of the Eleventh Edition of the *Encyclopedia Britannica*. It was eventually passed to me and I used and prized it during my graduate studies in English. The articles on English literature in the Eleventh Edition were then considered by some of my professors to be the best that were available at that time.

Homer Rhodes began life as a poor boy and when he could afford to, he bought things. He had at least one shotgun, a pistol, and two hunting coats. One of his purchases was a pony for his children. (Photo 24) Emory Holic, the blacksmith for the Elk Lick coal company, made a sled for the pony to pull. The pony balked and

Photo 24—Dorothy on the Pony

threw us out of the sled into the snow. We have pictures
of the gentle looking pony that should have warmed any

child's heart, but I for one was glad that mother prevailed and Daddy sold the beast.

Contemporary pictures show Homer Rhodes spent money dressing his family and himself--at least for formal pictures. (Photo 20 also 28) And even though we appear in some of mother's snapshots as dirty little brats I doubt if there was a ragged garment on one of his offspring at Spruce Knob.

On my seventh birthday, April 17, 1922, I had my father to myself most of the day. It was the day I discovered a loose tooth and he took me to Richwood to have it pulled.

We rode the cable car to the bottom of the hill. He and the brakeman on the train allowed me to climb up into the lookout box on top of the caboose for the ride to Richwood—I could see the front of the train where the engineer and fireman worked.

After we got to Richwood, Daddy took me to office of the Cherry River Boom and Lumber Company to show me to his bosses: Captain Armstrong, Mr. Hunter, and Mr. Baggett.

Before we went to the dentist, he took me into a drug store and bought each of us a dish of vanilla ice cream--served in the same kind of short stemmed pewter cup that is occasionally used today.

"Did you hear what happened to a man who came in here the other day and bought a dish of your ice cream?" Daddy asked the man behind the counter.

"Don't think so."

"Well," Daddy continued, "he bought a dish of vanilla like this one and started to eat it. Just then someone outside hollered 'Fire' and he left the ice cream and run outside to see where it was. Didn't see anything and come back in and found some one had eaten his ice cream."

"No kiddin" the counter man said.

"So he ordered another dish of ice cream and started eating it, when he heard a lot of people hollering 'Fire, Fire.' He run outside and the same thing happened. Couldn't see any fire, and come back in and his ice cream had been eaten."

"Don't believe it," the man said, "been no fire in Richwood for a month or more."

"Could of been longer than that", Daddy said. "The man, I forget his name, works with me up at the mines above North Bend. Anyway he ordered another dish and started to eat it when he heard a fire truck ringing its bell. This man wrote a note and stuck it on his ice cream with a toothpick. It said: 'Don't eat this, I spit on it!'

"Then he run out and watched the fire truck until it went by. Since he couldn't see any smoke or flames, he come back in and set down to eat his ice cream. Then he looked at the note. He noticed under where he had written: "Don't eat this, I spit on it!' someone had printed in big letters: 'SO DID I.'"

The man looked at Daddy sort of funny like, then he slapped his hand on the counter: "You rascal you, been pulling my leg all this time."

Daddy grinned, "Could it have been a drug store over in Summersville?"

"That's right, they had a big fire over there bout three weeks ago," the man at the counter said.

Then they both laughed.

"This is my big boy," Daddy said pointing to me, "gonna get his tooth pulled."

"Don't you go bawling," the man warned me. "Your Daddy won't let them hurt you."

I said, "I won't--I'm no cry baby."

We went and had my tooth pulled. It didn't hurt much and I didn't cry.

Daddy and I had one other time alone together. He took me to the First Methodist Church in Richwood to hear that great Democratic presidential candidate, William Jennings Bryan, talk about The Coming of the Prince of Peace. I believe Mr. Bryan was on his way to Dayton, Tennessee, to take part in the Scopes monkey trial, where he collapsed and died following the trial. After he was in Richwood, I read a notice on the front window of the newspaper office about Mr. Bryan's death.

Homer Rhodes got much more pleasure out of his children than his wife did. He liked us and enjoyed playing with us. Blanche's life centered about her husband until he died. And only then did her attention and devotion shift to her children--that is, HIS children. Of course she loved us, sacrificed for us, fought for us. But we were not the joyous whole of her being that was Homer Rhodes. We were just all that she had left of Homer Rhodes.

Age and memories can assist in understanding. Looking at the pictures of our parents in Dorothy's Album, at the photographs of the world they created together from 1910 to 1925, I can grasp the joyless void into which our mother fell when Homer Rhodes died. I believe that the most important thing in the life of my father and the life of my mother was his devotion to Blanche Rhodes and her devotion to him.

Years after their world ended, I also learned things about my father that I had never quite realized. While I was a student at the University of West Virginia in Morgantown, I hitched a ride on an A&P delivery truck to Richwood to visit Carrie Meyers. While I was with her in 1935 the radio program was interrupted to announce that Huey Long had been shot to death.

When I returned to Morgantown, I brought back the set of the Eleventh Edition of the *Encyclopedia Britannica* that our father had purchased for his family and for himself. I also brought back his mining books and a few of his old copies of *Coal Age,* as well as his red hunting coat, a pistol, and his straight razor. Except for a then out-dated Atwater Kent radio, the items I took were all that remained of Homer's possessions left by mother at Carrie Meyers' house in Richwood.

I wore out the red hunting coat, but when I cut myself twice with my father's straight razor, I gave it up for a safety razor. I gave the pistol to my son Stanley after he became a police officer in Norfolk, VA, and I passed the *Encyclopedia Britannica* to my other son Lloyd when I retired from teaching.

As a child under ten I had only a few ideas about my father. I believed that I had lost the most wonderful father in the world when he died. He was seldom angry. He never struck us or even touched us in anger. He bought us a pony and gave us things. He talked to us, played with us, and liked us. We found out he was really Santa Claus when we caught him trying to sneak a sled under the tree late Christmas night. We knew he was the biggest man on the hill—the boss of the mines. An idealized figure emerges in my memories of Homer at Spruce Knob.

By the time I left Mooseheart I had learned from mother that my father dropped out of the third grade when he was twelve years old to work in the coal mines. He was not quite twenty-one when they married, schooled in little more than swinging a pick, loading coal and looking at the back side of a mule hauling a coal car. At thirty-one, he was superintendent of the Elk Lick Coal Company mines at Spruce Knob.

I found in my father's books that I took when I visited Carrie in Richwood sample copies of the examination given by West Virginia to qualify a person to become superintendent of a coal mine. I realized then that Homer Rhodes had to have been a hard working-man, competent, driven by ambition, confident of his abilities, and very intelligent to pass the rigorous examination or take it.

But for me the most important thing I learned about my father was that their devotion each to the other was on based on his recognition of the fact that they were equals. Mother attests to this fact in her Little Book--a journal she started on our way to Mooseheart. She is in the hospital at Mooseheart on October 8, 1926 when she writes a love letter to her dead husband:

> "Dear Pal of Mine: -- Nineteen months ago today you were still with me, and I hoping against hope that you would be spared. . . . Our love was such a beautiful thing. Darling, the trials of life meant nothing to us so long we shared them. I love to remember the love light in your eyes as you replied to that lady's query about who was "boss" at our house. How

true your answer was – "there's no boss at
our house. We're <u>just pals.</u>"

I am fortunate and proud to be the son of Homer
Rhodes.

CARRIE THOMAS MEYERS

BLANCHE WAS PREGNANT with her first child (Shirley Lillian, who died at birth) when she opened the front door for a woman who said she was selling sewing machines. Granny Pritt was visiting her daughter in Cedar Grove when it happened. Granny often repeated the story about the second time she saw Blanche's birth mother by asserting: "I knew it from the minute I laid eyes on her—that it was her."

The woman supposedly selling sewing machines identified herself as Blanche's birth mother. She told Blanche and Granny Pritt that she was Carrie Euraine Thomas. She had been an unmarried school teacher when Blanche was born. Carrie said her own father was John Thomas, a blacksmith working on the Burr farm in Pocahontas County. Carrie told them that Blanche was born on July 30, 1891, in Poca, West Virginia. On a state map Poca appears as an unincorporated village less than twenty miles northwest of Charleston.

Neither Granny Pritt, nor my mother, nor Carrie Thomas ever offered me any information about how the birth mother was able to keep track of the child, then find her again, after having abandoned her more than twenty years earlier. Neither Granny Pritt nor Blanche ever gave

Dorothy or me any information about Blanche's father—
I doubt if they had any.

When I visited Carrie in Richwood, West Virginia, in
1934, while I was in college, she told me that my mother's
father was Henry Silas Burr. Carrie spoke wistfully of
Burr as a wonderful man with strong arms, the only man
she ever loved. That was all that she ever told me about
my mother's father. I wished I had asked her if he was the
owner of the farm where her father worked or was one of
the owner's sons.

Dorothy and her husband Earl Guinn have visited
the grave of Henry Burr, in a cemetery near Watoga State
Park in Pocahontas County. We have no picture our
mother's putative father. Burr's grave marker shows that
he was born April 10, 1865, and died March 23, 1937.
Dorothy believes the cemetery is on land that Burr may
have owned and farmed much of his life.

Carrie's reappearance was marked by an aggressive
campaign to take an active role in her daughter Blanche's
life. Dorothy found among our mother's papers the deeds
to three adjoining building lots in Cedar Grove, West
Virginia, purchased by either Carrie Thomas or Blanche
Rhodes.

The deeds suggest that Carrie purchased the lots for
Blanche or loaned her daughter the money to purchase
them. Obviously, Carrie's generous help strengthened
Carrie's influence over Blanche. The transactions contained the provision that in case of Blanche's death the
property "evades" back to the said C. E. Thomas. Our
mother outlived Carrie and the Cedar Grove property
would be of lasting importance to Blanche.

While Blanche was at Mooseheart, she gave her

mother-in-law, Annie E. Rhodes, a lifetime lease to the use of the property. Blanche was to pay the taxes while the property and any improvements on it were to revert back to Blanche upon the death of Annie Rhodes. Apparently Homer's father Isaac Rhodes had died after building two small houses on the lots. Grandpaw Rhodes and his wife Annie were living in the larger of the two buildings when we visited them in Cedar Grove to say goodbye before traveling to Mooseheart in 1926.

Beginning late in 1939, either Blanche or Marcena Pritt would live in the larger of the two houses on the Cedar Grove lots until 1965 when Blanche moved to Centralia, Washington, to live with Dorothy and Earl Guinn. Blanche assigned the three lots that Carrie purchased (or helped her daughter purchase) to George Cook, a Funeral Director in Cedar Grove, to cover the cost of her funeral; they passed to him when Blanche died on November 16, 1966.

Carrie Thomas's campaign for a role in Blanche's life created lasting tension. As Carrie and Granny Pritt struggled for influence and control over "their daughter" they also troubled the lives of Blanche and her children.

When our family stopped in Richwood, West Virginia, with Carrie Thomas Meyers they had a picture made on July 10, 1919. From there we moved on to Daddy's new job as foreman of the Elk Lick mines at Spruce knob. Blanche and Homer Rhodes were being drawn into her birth mother's world—the Elk Lick mines were about ten miles up the Cherry River from Richwood. (Photo 28)

George Meyers, influenced by his wife, Carrie, must have had a hand in our move to Spruce Knob—I suspect it was considerable. He was an official of some kind with

Photo 28—Bound for Spruce Knob

the paper mill in Richwood and kept track of the train loads of pulp wood for the mill. I know that he went away for several days to estimate a stand of timber for the mill somewhere over on Williams River. My father had gone there a year or two earlier to estimate the potential of a seam of coal for the Cherry River Boom and Lumber Company which operated the mines at Elk Lick. The paper mill and lumber company seem to have been connected in some way.

I doubt if Blanche ever loved or even liked Carrie. But she admired her and felt bound to her birth mother by a sense of duty. Blanche writes in her Memory Book on

> Sunday May 10, 1942: "This is Mother's Day. I feel so rotten. I didn't even send M. Carrie a card. However, I did write her a letter today. . . . Mother has been poorly all day."

The mother doing so poorly is Marcena Pritt. Blanche was then living with her and caring for her in Cedar Grove.

Blanche's Memory Book also records that she was forced at least once to choose to live with and take care of the woman who took her in as a baby and thus to reject her birth mother. Blanche notes in August 1947

> I was confronted with a problem that almost drove me to distraction for awhile. Mother Carrie wanted to come here to live. It would have given me peace of mind, and been to my advantage in a financial way. I went to Richwood, and we planned it all out. I came home and told

Mother Pritt. She said she would leave if
Carrie came, so the subject was closed."

I suspect Blanche was awed by the fact that Carrie
had been a school teacher and owned property. I sensed
that she admired Carrie as a go-getter filled with ideas of
how to succeed in the world. For a time, Carrie used the
largest room in the front of her house at 16 Riverside
Drive as a grocery store. (Photo 29) Years later she bought
the Richardson place next to her house, probably for a
grocery store. Neither venture seems to have panned out,
but Carrie somehow found the money to start them.

Dorothy believes that our father, Homer Rhodes,
liked Carrie more than Blanche did. The move from Kanawha county to Spruce Knob was a good one for him:
as a foreman and then superintendent of a coal mine he
earned much more money than he earned swinging a
pick and loading coal.

Economically the move helped our family, especially
after the death of Homer. Charlie and Granny Pritt could
not have supported us as Carrie and George Meyers did
for almost eighteen months—between the time when
Homer became ill and our move to Mooseheart.

Yet the long term emotional effect of moving to
Spruce Knob and bringing Granny Pritt to live with
us was bad. The struggle between the two mothers for
control of their daughter and her family intensified.
Blanche began to address Carrie Meyers as Mother and
refer to her as Mother Carrie. Blanche's children found
themselves with a Granny Carrie and a Daddy George.
Marcena Pritt was always called Mother by her daughter

Photo 29 – 16 Riverside Drive, Richwood, WV.

and Granny Pritt by Blanche's children. Every one called Charles Pritt just Charlie.

When we moved to Spruce Knob, Carrie began complaining about Granny Pritt's partiality to toward me; she labeled me as Granny Pritt's boy and announced that Cliff was her boy. She fastened upon the normal sibling rivalry between Clifford and me and turned it into competition and jealously that seeped into our early years.

Our rivalry developed into childish fist fights that continued until I left Mooseheart in 1933. Our rivalry kept us from developing the kind of tolerance, admiration and affection we both shared with our beloved younger brother Clyde. It kept us from feeling for each other the kind of brotherly love that we both shared with Earl Guinn—our adopted brother. Earl married our only sister Dorothy whom Cliff and I both idolized. She of course loved both of us equally and perhaps more than either of us deserved.

The rivalry put almost as much emotional distance between Cliff and me as the decades and the three thousand miles we lived apart. He lived the greater part of his life in California, while I have always lived east of the Mississippi. We rarely exchanged letters, but for short time before he died we shared family news through E-Mail and he forwarded reams of West-Coast humor—that I detested.

I have a granddaughter by marriage, another by adoption, three by ties of blood, two great granddaughters and a great grandson. I will never admit that I have a favorite. If by circumstances one should become my favorite, I will deny it! Both of my Grannies taught me never to turn children against each other by playing favorites.

After my father died, I remained at Spruce Knob with Granny Pritt to finish school. There I had her full attention and sympathy to master the loss of my father—to the extent that child can realize the death of a parent. When school ended at Spruce Knob and Granny moved back to Mammoth, then I moved to 16 Riverside Drive in Richwood to be with the rest of our family and Carrie Thomas and Daddy George. My life was miserable for the next sixteen months.

I first became aware of being an object of Carrie's antagonism by accident at breakfast the day after moved from Spruce Knob to live with our family in Richwood. I didn't like the oatmeal Carrie had fixed, and said that "Granny Pritt cooks oat meal in a skillet."

"Granny Pritt's not feeding you—we are. You'll eat what we put before you and like it."

I ate! But I didn't like it.

A few days later she bought Cliff a bicycle, a real two

wheeled bicycle. She insisted that I had to ask him for permission when I wanted to ride it. I learned how to ride the damned thing, but I didn't learn much about sharing or generosity.

Carrie was stout, frumpy, bossy, and always smelling of Sloan's liniment. She slept with a block of wood under her head, because "that's the way the Chinese do it and it is good for you." I never liked her.

Carrie had a streak of cruelty. She was the only person I ever heard of who forced people to take castor oil as a "punishment"—except the infamous Mussolini. She made us take one or two tablespoons full depending upon the seriousness of the infraction—or one of her whims. It was Carrie's theory was that children who misbehaved were infected with badness that could be cleaned out through their bowels.

I believe her theory of purging badness from children derived from the treatment for intestinal worms that she inflicted on us during the two springs we were subjected to her care. The treatment consisted of swallowing calomel powder wrapped in paper followed by a pint of mineral oil. We would have preferred the worms to the treatment—if indeed we had worms. But whether it was worms or badness the treatment was mineral oil or castor oil—depending on which she had on the kitchen shelf.

Dorothy told me she was always afraid of Carrie who whipped her bare legs with a thin switch. Dorothy charged that "She pinched me to make it hurt. She made me memorize Bible verses, go to Sunday school and recite them."

I became the new kid in the sixth grade at Richwood where I made only one friend a boy named Deward Booher. Once during a snow storm we went out in the woods on the hill behind Richwood to get a Christmas tree for the classroom. We found a tree, but it is still a wonder to me that we didn't get lost and freeze to death.

The teacher was a man I don't remember well enough to say that I liked or disliked him. The little girl that I wanted to know, Mamie, ran with the gang from across the river. They were usually busy passing notes to each other and moving around the room whispering: "I will if you will!" I sensed that they were talking about the kind of fun that Daddy was going to explain to me when I was old enough to enjoy it. Apparently I wasn't old enough.

We had chores to do at 16 Riverside Drive. Cliff and I had to gather scraps of wood and bark from the bottom of empty pulp wood cars and carry it home in bushel baskets—kindling to start fires in Carrie's kitchen stove. But more than that, we were expected to watch for the pulp wood cars and gather the kindling before the empties were pulled away to be reloaded. And we weren't even paid for the work; at Spruce Knob our Daddy always paid us for doing a man's work—like carrying in wood.

One of the few times I was ever whipped was by Daddy George for failing to collect sticks and bark from the pulp cars. Both Cliff and I shared the irritated blows from a long buggy whip. Daddy George never had a horse or a buggy. He bought the whip after Blanche's brats were dumped on him to be fed and disciplined. When we failed to do something we were ordered to do, he had to whip us into obedience—"spare the rod and spoil the child."

I will always remember how he pushed us into the room where he stored feed for the cow and chickens he kept under the house. Shoving us together side by side, he made us take down our pants, bend over and present our bare butts for whipping. Then he took a couple of half hearted swings with absolutely no intention of inflicting pain or raising a welt, ordered us to get out, do what we were supposed to do—when we were told to do it.

Daddy George didn't have Carrie's cruelty. He was more snarl than bite. I don't believe that he ever struck us again, but that one whipping was an insult to our pride. It cut as deeply as his constant reminder that:"I'm a feedin' ye, ain't I?" It told us he didn't want us, he didn't like us, and he had no idea of what to do with us.

I had the job each evening of going out to find the cow by the sound of her bell and bring her back from across the river—driving her under the house to be fed and milked. I also had the job of tending to Daddy George's patch, a small vegetable garden, on the side of the hill, weeding it and bringing home radishes, lettuce and spring onions.

A part of my job at Daddy George's patch was to take rabbits from the box-trap, bait it again with a piece of apple, and reset it. I always had to kill the rabbit, a job I didn't like. The rabbit sometimes squealed when I held it by its hind legs to hit it on the back of the neck with a stick. Then I had to hold it by its head and swing it around until its head came off. I didn't mind killing chickens by wringing their head off—they didn't squeal. We often ate fried rabbit, which was not as good as chicken.

Daddy George never paid us or praised us for our work as our father always did. With little money and no way to earn it, I became a sneak thief—robbing my piggy bank of the dimes my mother gave me to save—spending them at Cherry's Drug store for jawbreakers, chewing gum, and Wings cigarettes.

In fairness, Daddy George was generally a good man. Mother and Dorothy both liked him, trusted him, and he liked them. He had no reason to like Blanche's three boys, to take care of us or feed us.

Daddy George was on the Board of Directors of the Citizen's Bank when it failed during the depression. The bank failure tied up $30 I got for Daisy's calf and $32.50 more I had earned for building the fire in the schoolhouse stove at Spruce Knob each morning before classes started. Years later the bank repaid all of it, while I was in college at Morgantown in 1935, sending me five checks for $12.50--one each to be cashed on the first day of five successive months. Those checks paid most of my tuition, $75 for the last two semesters of college. Mother told me that Daddy George and some of the other members of the Board used their own money to pay off the depositors like me. Daddy George was basically a good man—but he didn't like us, so we didn't like him.

One morning Carrie was puttering around and cornered me in the upstairs of her house where we slept. She bawled me out for being sloppy, and showed me how to fold my towel neatly over the rod on the marble-topped washstand.

That encounter etched into a sloppy boy a compulsion for a certain kind of neatness. I believed it would make people in authority such as Carrie like me. In any case I always arranged my possessions neatly in the metal lockers I used at Mooseheart, wanting to please the matron or proctor. I doubt it they noticed, but it became a habit with me.

Carrie was a careful teacher; for example she showed me how to fold my letters so the person receiving them would see the salutation when they opened them. Carrie had a strong influence on me. Years later, I occasionally taught Bibliography and Research Methods to Graduate students in English—and enjoyed it. To this day, I sort the books on my shelves by subject and then arrange them by height—to the amusement of my kin and friends.

It was unfortunate that Carrie Thomas failed herself and me. I could have learned to like her, had she merely pretended to like me. But by the time she died in 1956, I was still too immature to recognize the importance of Carrie's kinship and her contributions to my totality.

I often lived within driving distance of Carrie after I left Mooseheart. But I saw her only twice before she died: once while I was in college at Morgantown and once when I went with Blanche to visit her. My blood grandmother never wrote me a letter—nor even sent me a post card, as far as I can remember. We never exchanged gifts or favors of any kind. For Carrie Thomas and her eldest grandson, it was always too early or too late for understanding, or affection, or unvoiced forgiveness for what Ibsen has sometimes dramatized as the crime of the denial of love.

SPRUCE KNOB

MY SONS, STANLEY and Lloyd, frequently went me to Richwood, WV in the l980s through the l990s. We drove past Carrie and George Meyers' house at 16 Riverside Drive and sometimes took a picture of it. We have seen the old wooden building in various stages of deterioration and repair, but it doesn't look much different than it did when our family was trapped there for almost eighteen month in 1925 and 1926. We usually visited the graves of my father and mother in Plot 107 in the cemetery on a hill outside of Richwood. Carrie Thomas and George Meyers are buried in the same plot. In this way, Carrie Thomas finally triumphed over Marcena Pritt in the struggle for their daughter: Granny Pritt is buried at Ward, West Virginia, near her second husband, Charles Pritt.

On our several trips to Richwood, we searched the side of Fork Mountain for our home at Elk Lick, a coal mining community that we usually called Spruce Knob, which has disappeared into the forest-size trees and underbrush of the Monongahela National Forest. We made several trips before we located the site of the home of Blanche and Homer Rhodes.

Finally we found a large split rock and I knew we

were at Spruce Knob. Dorothy's Album includes a picture showing Clifford and me sitting in front of that rock between our playmates John and Joe Postek. (Photo 25-- Kids at the Split Rock) Another picture (now lost) was made of me standing in the crevice of the same split rock. I remember the date on the back of the lost photo: March 9, 1924. The lost picture and Photo at Split Rock were made on the same day—the day that Emory Holic had to kill Old Bess, a mule who broke her leg in working in the Mine. Homer Rhodes died one year later On March 9, 1925.

Emory was the blacksmith and general handyman at the mines. He allowed me to follow him around and watch him work. Emory was my favorite of the people who worked for my father. I think of him as another of my teachers—and along with Granny Pritt, one of the very best.

Daddy told mother that Old Bess was injured on the last day of work scheduled in Opening #7. If she had not broken her leg, she would have been retired because the other openings were too low for her. Company policy required that mules with broken legs be destroyed for humane reasons.

I went with Emory when he killed Old Bess. Carrying a sledge hammer, he led the hobbling mule out of the barn to a place beyond the split rock. There, away from the houses, Emory swung the hammer and the mule dodged—jerked her head away. Emory then put his blue checked handkerchief across her eyes and smashed her to death with a single blow on the place where the handkerchief lay. She crumpled and rolled down the hill. Later he would pile wood about her body and burn it—as he

Photo 25-- Kids at the Split Rock, March 9, 1924.

did with other mules destroyed following injuries in the mines. Several piles of burned wood marked and white bones marked those places. I did not stay to see Old Bess' body burned up.

The killing of Old Bess when I was nearly nine years old was my first encounter with death. The second was the death of my father a year later. His life of thirty- five years and six months ended in a Baltimore Hospital on March 9, 1925.

My sons and I did find the site of the main entry to the mine. I knew the entry into the side of the mountain was a short distance around the hill from the split rock. And the Forest Ranger in Richwood explained to me that years before the entry had been blown up and fallen in upon itself, it left a depression effectively blocking adventurers from losing themselves in the cave-like halls hacked out of the earth by miners.

The Ranger said that a logging road runs over the original grade cut for the motor that hauled coal from the main entry to the tipple. The original grade wound around the mountain some 1200 feet above the bed of the Cherry River and continued from the remains of that tipple to another opening known in my father's day as the new entry. It had also been blown up and fallen in upon itself.

Stanley and Lloyd and I followed the logging road from the main entry to a cleared place on our right hand that lay down the hillside littered with pots and pans of various shapes and sizes. We noticed a few fireplace bricks scattered in the litter. I was home!

At least I was at the site, in reference to the main entry, where superintendent's house had been. And that house, the earliest home that I clearly remember, was the only house at Spruce Knob which had a fireplace.

We carried several of the bricks back to Norfolk, where two of them served as doorstops in my house until we left it in September 2003 following the flood during Hurricane Isabel. Our discovery of the bricks set us looking for someone to help piece together more about Spruce Knob and the people who worked there.

A year after we located the site of the superintendent's house and discovered the fireplace bricks, we found the monitor's track leading up from the lower tipple at the bed of the Cherry River to the remains of the tipple on the logging road. An old-timer who lived outside of Richwood showed us the scar on the side of the mountain made by the monitor's track. We waded across the Cherry River at a shallow place about a hundred yards toward Marlinton from the "10 Mile Branch" road sign.

It was getting dark, but the track was marked by pieces of coal and slate that had fallen from the monitors making it distinctly different from the woods on each side of it.

The trees and undergrowth had not reclaimed all the land. We found spikes and pieces of cable at the place where the two monitors passed each other—the loaded car going down the hill and pulling the empty one up to the tipple to be loaded. However, when we reached this passing place it was too dark to continue up the hill to look for the remains of the tipple on the logging road.

In May 1991 we found someone to help us locate the sites of several houses and other structures connected with the mine at Spruce Knob. By chance we met John Postek at the Fire Department in Richwood. John had been one of our playmates at Spruce Knob, and is shown with Cliff and me and Joe Postek (Photo 25) at Split Rock. John suggested that his brother Joe, of Barberton, Ohio, might go with us to the abandoned mine site.

John and Joe, with three younger brothers, Stephen, Andy, and Paul, lived next door to us at Spruce Knob with their parents John and Agnes Postek. John told us they stayed at the mines after our father died. John said Joe, in particular, had kept his interest in Spruce Knob and its people.

I wrote to Joe Postek. He agreed to meet me and my sons in Richwood and go with us to the mines. Joe also sent me a Xeroxed copy of a clipping from the Richwood News Leader (dated 1971 in the margin—I suppose by Joe). The clipping contained the names of the people Joe remembered at Spruce Knob. Some names I had not heard or seen in print since 1925 popped vividly into my mind, names without faces, and sadly names that sink again into the dark as my eye passes. The newspaper clipping reads:

HERE FOR FESTIVAL

Former Richwooder, Joe Postek, and son David 10, of Barberton, Ohio, were in for Homecoming and added more info to the CRB and L' s [Cherry River Boom and Lumber Company' s] North Bend operation featured in a recent News Leader. Joe went to school there and remembered his teachers, Capitola Cogar and Fay Chapman. Residents he remembers were: John Overbaugh and his sons (the late Julius Overbaugh was janitor at the school). T. Cunningham,

G. Nichols, Ford Sergeant, Bert Harris, Tom Brewster, George Jackson, Lon Spencer, [Charles] Homer Rhodes, John Cross, Emory Holic, Andy Husark, and sons Steve and Andy, Jr., Bunt Hardman, George Whitlatch, Brightman Young and George and Jack, John Cruchshank, George Tawney, Bob Haynes [husband of Eva, our mother's first cousin, and father of Toots and Beulah Haynes—our playmates at the mines and later in Richwood], Jack Dellhoney, Joe Boushotes and Louis Prelaz. Mr. Postek is visiting his brother, John, of Southside. He reports that his mother [Agnes] left here several years ago and lives next door to him in Barberton. She is 83 and enjoys good health.

Charles Pritt, Granny Pritt's husband was not on Joe's list. Also missing was Mr. Williams who lived with his family in a house near the lower tipple at the bottom of Fork Mountain. I believe one of Williams' jobs was to see that railway cars were in position to receive coal from the monitors. The family included two boys who sometimes threw lumps of coal at Clifford and me to keep us from the apple trees at bottom of the hill—they considered the trees theirs. I remember also that Mrs. Williams had twins while we were at Spruce Knob-one of them died of diphtheria shortly after birth.

A third person missing from Joe's list was an unmarried miner named Miller, the brakeman on the motor.

He was the only person I remember as killed at the mines while we were there. Daddy told mother that Miller jumped from the motor and threw the switch to shunt the coal cars onto the siding and into the tipple. The loaded cars gained momentum as they rolled down the siding. Miller's jacket caught on one of the cars, he stumbled and was dragged under the wheels and killed.

When we arrived at the split rock Joe and I began working our memories to recreate a picture of our childhood home and identify the people at Spruce Knob. He confirmed that the Forest Ranger was correct about the logging road following the grade of the motor track linking the two entries of the mines and the tipple; Joe agreed the scar on the side of Fork Mountain marked the track of the monitors which carried coal down the side of the mountain to be taken to the lumber mills in Richwood.

Our childhood playmate told us that after my father died Lon Spenser became superintendent of the mines and John Cruickshank was made foreman. Joe was not exactly sure when the coal ran out and the mines closed—years before World War II began. He offered the theory that the motor track rails were torn out, scrapped with other heavy mining equipment, sold to the Japanese and blasted back at him while he was with the Marines in the South Pacific. If so, the scraps were also blasted at me while I was stationed at the Lunga Point Boat Pool on Guadalcanal from January 4, to the middle of June 1943. In any case, work stopped, the mines closed, the wooden houses were knocked down, carried away, or burned before World War II. Trees grew where the miner's houses had been. The trees matured, were cut down, and logs carried away on trucks over the road once used to haul coal from inside the mountain.

Joe showed us a stone reservoir between the split rock and the main entry. It was filled with clear cold water from a mountain spring, overflowing down the side of the hill, as it did in the early 1920s. Joe believed it provided water for the boilers at the power house that ran the steam generators producing electricity for the machinery and houses at the mines.

The reservoir is fixed in my memory as the one thing at Spruce Knob that seems untouched by time. In freezing weather Daddy would send Clifford and me with our sled to gather ice that formed around the edges of the reservoir and the entry to the mine. More than once we hauled ice home for Daddy and Mother to make ice cream.

Joe told us that a short distance up the hill, behind the reservoir, and to the left of it had stood the powder house, in which blasting powder, fuses, carbide for the miner's lamps, and similar materials were stored. The power house had a particular part in my life at the mines. It was my regular job to drag the empty powder boxes home and split them up for kindling to start the coal burning in the kitchen stove and in our fireplace.

The mule barn was farther up the hill behind the powder house. Joe said that near the barn there was once a small house in which Emory Holic lived. The trees and brush of Fork Mountain have erased the sites of all of those buildings

Joe told us that Emory was his uncle, that he had a noticeable limp, and that Emory's wife died during childbirth. I remember none of those things, and felt a passing tinge of jealously about Joe who must have been closer to my early teacher than me.

Emory was the blacksmith; he shoed the mules and sharpened the bits which fastened into a machine that cut into the face of coal seams. Emory's blacksmith shop was near the power house. Sometimes he let me crank the bellows; operated by a geared system, the bellows forced air into the glowing coals turning the mule shoes cherry red. He would lift the shoe with tongs to his anvil and pound the hot piece of metal in an explosion of sparks. Then he cooled the shoe—hissing in a pan of water. I worried that the metal, which appeared to still be hot, would hurt the mule as Emory nailed the shoe to its hoof; but the animal didn't seem to mind.

The power house site down the hill, not far from the main entry, was marked by rusted pipes, the fire box for the boilers, cement piers on which the boilers rested, as well as the grade down which cars carried coal to fire the boilers. Joe believes that most of the pipes that carried the water downhill from the reservoir were also cut up and sold to the Japanese for scrap.

Although Emory was the blacksmith and took care of the mule barn, I think his main job was to run the power house. Emory checked the large clock-faced gages and responded to their demand for coal and water; he tended the humming generators which kept power flowing into the copper line carrying electricity for the houses and for the operation of machinery at the mines—especially the motor that pulled the coal cars to the tipple. My sons found a two inch length of a ¼ inch copper wire, which Joe identified as a piece of the power line. Dorothy is shown on the pony on the motor track below the naked power line (Photo 24). And another picture (Photo 27) shows that the motor worked like a street car, drawing power through a pole from a naked line above the track.

Joe and I could not find the slate pile located near the power house and black- smith shop. Emory never allowed us to play on it. He warned us that the slate could start sliding down the side of the mountain and bury little boys.

We continued along the logging road from the power house and blacksmith shop until we came to the place where Stanley, Lloyd and I found the bricks from the fireplace in the superintendent's house. Joe confirmed that we were at the site of what was once my home. But like me he was unable to explain the presence of so many tubs and pots and pans as well as the absence of other kinds of litter—bottles and jars and similar trash. But he was certain that we had located the site of four houses that were at Spruce Knob from 1919 until after Homer Rhodes died in 1925.

Joe said Mr. and Mrs. John Cruchshank lived in the house closest to the blacksmith shop. We both recalled how Mrs. Cruchshank used to make seafoam candy and sometimes sell it on Friday afternoons for a nickel a bag.

The Posteks lived in the house next to Cruchshank's house and we lived in the house with the fireplace, next to Joe's house. The Prelaz family lived in the house on the other side of us.

Joe found a small spring-fed reservoir up the hill about equally distant from his house and ours. He insists that, while our house had running water, his mother had to walk up the hill to get water for her family. I am not sure that he is correct, but I know that Mrs. Postek made the most wonderful doughnuts by frying the dough for "light bread" in deep grease. Then she rolled the hot doughnuts in white sugar. She taught me the words in Czech for bread and apples which I remember (phonetically) as "klebi" and "yahbuco." Joe Postek and his brother John were our regular playmates, along with our cousins Toots and Boo Haynes, the children of Eva Haynes, daughter of mother's Aunt Sue Thomas. They are present (though it is difficult to identify them) in many of the pictures of children at Spruce Knob. Joe, who is a few years younger than me, remembers our sister, Dorothy, as his partner in making mud pies.

At one side of the reservoir from which our families drew water was a scum-covered pond originally filled with frogs. There baby brother Clyde, born at Spruce Knob on December 31, 1920, grew old enough to collect tadpoles in glass canning jars. That activity earned him the nickname, Tad, which he carried until we left the mines.

A coal house and an office for Daddy stood between the in motor track and our house. Daddy kept blueprints in his office showing the working openings of the mines. Photo 23 shows Homer in the door of his office with Ernie, Cliff, Joe Postek and Dorothy. We are standing on the wooden walk that stretched from the coal house beside the motor tracks to the office and on to the door to the kitchen of our house—raised on stilts on the side of the hill. Underneath the house we kept chickens and had a pen for the cow, Daisy, as well as a room in which we kept feed for the animals. Down the hill from the house was a two-seated privy.

Mother cooked our food; we ate it and lived in the

same room. A kitchen stove stood beneath the windows in the wall opposite the door. We ate our meals at a round dining table that stood on a pedestal not far from the kitchen stove. A fireplace warmed the room from the wall at the right of the door through which we entered. There in front of the fireplace we were subjected to a weekly bath—on Friday or Saturday nights. The tub was the same oblong vessel that mother placed on the coal burning kitchen stove to boil water for washing clothes on weekdays—usually Mondays.

Later Daddy installed an Atwater Kent radio in the room. We could get a couple of stations on the radio: KDKA, and WLW. Outside the window we could see the antenna strung from the house to a large pine tree.

Below our house, between the pine tree and the privy, Daddy hacked out a garden for vegetables. We used the stems of green onions from the garden as containers for Prince Albert tobacco filched from our father to make cigarettes. Didn't work well, so we tried other things to improvise cigarette paper including the outdated Sears and Roebuck catalogue kept in the privy as toilet paper. We also "smoked" pieces of a twig we called camphor vine, cut into cigarette lengths. It could be dried, ignited at one end, kept glowing by sucking on the other end and blowing the acrid smoke out of our mouths.

The Prelaz family lived on our left, as we looked down the side of the mountain. I once killed a red squirrel in a beech tree down the hill from the Prelaz's house. I hit the squirrel in the throat with a single shot from my BB gun, and it tumbled dead at my feet. I skinned it. Granny Pritt par-boiled it and fried it with flour gravy for me. The shooting happened after Daddy died while Granny and I lived together in our house at the mines. The first airplane I ever saw flew over the big pine between our house and the Prelaz's. A few days later it crashed and killed the pilot.

One of the earliest struggles with my conscience happened while Granny Pritt and I lived at Spruce Knob. I did something I knew was wrong. I was caught and I lied about it. Then, I attempted to erase my sense of shame and guilt.

The incident occurred because I was not allowed to fish with a barbed hook. Granny gave me a piece of white twine and a safety pin that I fashioned into a hook without a barb. I went down the mountain to Cherry River and dropped my homemade hook, baited with a worm into a dark hole at the edge of a big rock. Miracle of miracles: a fair sized trout took it immediately. I pulled the fish out of the water onto the rock. The trout flipped off the barb less hook and flopped back into the river. That was the last bite, the last bit of luck, I had on the river.

I finally gave in, and started up the mountain following a good sized feeder stream. Now, I knew that it was against the law to take fish from a feeder stream. And I knew it was against the law to keep a trout that did not measure at least six inches from the tip of its nose to the fork in its tail.

I saw a trout, chased it into in a shallow pool and caught it with my hands. The trout barely measured six inches from nose to tip of its tail—and possibly a half inch less. It wasn't a keeper. I broke two laws by taking an under-sized fish from a feeder stream.

I ran a string through the fish's mouth and out of its gills. I continued up the side of the mountain to the motor track proudly dangling my trophy. Lon Spenser, who now had my father's job as Superintendent, stopped me to look at my catch. He asked me where I caught it.

"By the big rock up from the lower tipple," I lied.

Lon looked at the fish and said, "Your Daddy would be ashamed of you for keeping an undersized fish." I wasn't sure Daddy would have been ashamed of me. Homer Rhodes was a meat-hunter rather than a strict adherent to West Virginia's hunting and fishing regulations.

Nevertheless, I was ashamed of by being caught with an undersized fish and of lying about where I had caught it. Consequently, I just wanted to get rid of it. Granny Pritt helped me bury the fish beside the kitchen door The earth was soft and it was easy to dig a hole to bury the fish because it was the place where Granny threw her dish water and coffee grounds.

It was about that time when I found a blue crawdad (crayfish) in the soft ground where we buried the fish. I asked a biologist at Old Dominion University about it years later. He was a specialist in West Virginia creatures like crayfish and he assured me that I hadn't dreamed the whole thing up. He said blue crayfish are not uncommon in West Virginia, and that he had seen them in the Cranberry River area. I didn't ask how the crayfish got into the wet spot half way up the mountain from the river where it belonged.

Joe Postek located the site of the schoolhouse on a slightly level spot beyond a pile of rocks on the upside of the original grade. There he had been a student through the third grade. I also completed several years there.

We found a spring and a small pool of water near the

site, where we got our drinking water during the days we were in class. Behind the schoolhouse and sloping up the hill, an open space ended in thick dark woods that ran toward the top of the mountain. I remember the woods beyond the open space as mysterious. I sometimes had to venture into them to find our cow Daisy near the top of the mountain—where the world ended. The search made me uneasy.

The schoolhouse was painted red, and had but one room and one teacher for all of the school-age children at the mines. A coatroom was partitioned off opposite the entrance. On the walls of the coatroom were pegs on which hung our coats and caps.

A round pot-bellied cast iron coal-burning stove vented by a black sheet metal stove pipe was centered in front of the coat room partition. I have no recollection of a desk for the teacher, or a blackboard, or windows in the schoolhouse.

I recall the stove, because for a time Daddy paid Emory Holick five dollars a month to build a fire in it on school days. Emory paid the money to me because I built the fire every morning before school started—regardless of the weather. I believe it was some kind of a scheme between Daddy and Emory to teach me responsibility.

The schoolroom was equipped with a dozen or more desks arranged in two rows with an aisle between the rows. Each desk accommodated two students and had an inkwell for each student. At one of those desks I suffered through penmanship using split steel-nibbed pens. And seated there, I also learned how to relax on command by the teacher. She made her students rest their heads on their folded arms and remain absolutely quiet for one full minute. Occasionally, I still do it.

In that school room I learned that the stag at eve had drunk his fill where danced the moon on Ronan's rill. I learned (gradually) what a stag was, that eve was another word for early evening and that a rill was a kind of a stream or pond. But I never quite understood how the moon could dance on it. In that school I met robin redbreast. There I was impressed by the story of the kindly lion that spared a mouse. That mouse later saved the lion from a great rope net by gnawing the net apart. Quite a mouse, I thought.

Joe remembers that Fay Chapman and Capitola Cogar taught there at different times. I don't remember Miss Chapman. Capitola Cogar once paddled me. I went home and told Granny Pritt that the teacher broke a corner of her big paddle on my behind.

"What did you do?" Granny wanted to know.

"Nothin,' she jist don't like me."

"She break that paddle on you?"

"Uh huh—"(with a teary whine)

"Fetch it!" Granny ordered.

I went back to the school house and took the paddle.

"That'll fix 'er," Granny growled, as she lifted the lid of the kitchen stove and we put the paddle on top of the coals.

Long after the wood had been reduced to ashes, the sculpted handle, with a hole for a throng, and the broken corner of the paddle were outlined clearly in the bed of coals. The image is etched sharply in my memory still— one of my special spots of time.

I have no idea of why Miss Cogar paddled me. But I'm sure it never did any of us any good. And that's that—as far as I am concerned about the effectiveness of corporal punishment.

One family kept pigs in a pen beside their privy down the hill, not far from the school house. Again, a spot of time: a pig being butchered. The animal was stunned by a blow to the head with a sledge hammer, and then stuck in the throat with a long knife. The pig struggled and squealed while the butchers held it and collected the blood that flowed from the pig's throat. The butchers explained the blood would be used to make blood pudding, but I never saw or ate any of the pudding.

Most people listed in the newspaper clipping that Joe sent me lived in houses on each side of the motor track between the schoolhouse and the tipple. Little remains of those houses except rusting pots and pans. The only picture I have of the miners' homes at Spruce Knob shows three buildings on the upper side of the motor track beyond the schoolhouse. They appear in the background of (Photo 26) showing Clifford and me with John and Joe Postek and the pony. Joe said perhaps five or ten more houses, including a boarding house, were located in that area on the downside of the motor track.

Neither Joe Postek nor I remember apple trees we found along the upper side of the motor track beyond the school house. Joe has the theory that hunters planted the apple trees to attract deer. In any case the twisted old trees were planted after the coal was taken from the mines and the wooden buildings torn down and carried away or burned.

Photo 26 —Miner's Houses behind the Kids and Pony.

One of our discoveries supports Joe's theory: up the hill behind the twisted old apple trees we saw a deer hunter's stand (a platform built in a tree) where a hunter sits and waits for deer to pass within range of his gun. Incidentally, we saw two deer move up the side of the mountain while we were there.

One or two apple trees did grow near the tipple at the bottom of the hill, while we were at Spruce Knob. The Williams boys who lived with their parents in a house at the bottom of the hill believed the apples belonged to them. They tried to stop Clifford and me from gathering green apples, chasing us back up the mountain, throwing lumps of coal at us. Nothing ever tasted as good as good as those fried green apples, with hot buttered biscuits. And surely, those little green apples were made for us— especially since mother encouraged us to get them and then fried them for us.

The concrete foundations of the tipple and the building housing the cable car mark the center of the mining operation at Elk Lick. A picture from Dorothy's Album (Photo 27) shows the cable car house in the background on the left and the tipple on the right. The motor in the foreground has just hauled several loaded coal cars from the mine. The motorman is at the controls but it is difficult to see him. Three men stand at the other end of the machine. The brakeman is on the left, Lon Spencer is in the center and Homer Rhodes is at the right of Lon.

Two unidentified workers push a loaded coal car into the tipple to be weighed. After the coal is weighed a numbered brass coin-like check will be lifted from the side of the car and the load credited to the miner identified by the number on the check.

Then the loaded car is pushed onto a hinged platform until it is stopped by a pair of curved rails—shaped like giant fish hooks. A pin is removed to allow the front end

Photo 27—Motor at the Tipple

of the car to swing open. The platform on which the car rests will be tipped up and the coal spilled into a large steel car called a monitor. Then the emptied coal car will be pushed onto the siding at the right of the picture and hauled back into the mine and filled again with coal.

There were two eight-ton monitors at Spruce Knob. When the monitor at the tipple on the side of the hill was loaded, it was allowed to roll down the single track and pulling the empty monitor up to the tipple at the top of the hill. At a point near the middle of the single track the monitors each moved onto switched tracks, passed and then moved back onto the single track. When the loaded monitor reached the tipple at the bottom of the hill it was dumped into a railroad car. The empty monitor, upon reaching the tipple on the side of the hill, was again filled with coal and the process repeated.

When the railroad cars at the bottom of the hill were filled they were hauled to Richwood, about ten miles away. The coal was used to power the lumber mill of the Cherry River Boom and Lumber Company in Richwood.

To the left of the tipple on the side of Fork Mountain (Photo 27) stands the cable car building. It housed an electric winch used to haul a car filled with supplies for the mines up the side of the mountain. We called the system "the incline" but mother called it the "the car" or "the cable car." The system included a very steep incline constructed of timber that arched up from the side of the mountain to the level of the building housing the winch. People often rode the car all the way up the incline but mother always refused; it was scary, even for some of her kids.

One day I met Emory Holic on the motor track pushing a coal car. It was raining and I was carrying an umbrella. "Get in boy and keep the rain off him," Emory ordered. I got in and saw Lon Spenser lying in the bottom of the car. He wasn't moving or saying anything and blood was coming from the side of his head. Emory said Lon had been using the motor to pull a loaded coal car. The cable between the motor and the car snapped in two and hit Lon.

Emory pushed the car with both Lon and me in it around the motor track to the cable car house. He lifted Lon into the cable car and took the umbrella from me. I was glad my job was over. I was afraid to ride down the incline because Mother always said it was dangerous.

Anyway, Emory didn't need me. I knew that when he got to the bottom of the mountain Mr. Williams or one of the men at the tipple at the bottom of the hill would help him get out the motor car for Richwood and start it.

It was an ordinary Ford automobile fitted with wheels to run on a railroad track.

All Emory had to do was push it out of its shed, jack it up, swing it around and set it on the tracks. Usually Daddy did the job without help. The hardest part was cranking and getting the motor car started.

There was not much talk about Lon's injury. But he sure looked hurt in the bottom of that coal car. Daddy must have been ill and away when Lon was injured; at that time Daddy was often away because of his stomach pains. Mother seemed to think it was from ulcers or perhaps an old injury—Daddy had been kicked in the stomach by a mule when he was a young man.

The building housing the winch and cable car burned mysteriously a short time after Lon Spencer was hurt. I remember how Mother and Daddy talked quietly so we couldn't hear what they were saying as we watched the building burn and followed a light moving down the side of the mountain. I always wondered if it was it carried by someone who set fire to the cable car building.

Father and mother said nothing to explain the fire or the light. Once I heard Daddy say something that could have been interpreted as recognition of hostility towards the coal company and its officials—including Homer Rhodes. During a shooting match, A miner suggested that Daddy post, and collect the targets, and help judge them. This would have put him down range from a group of men with loaded shotguns.

"Hell no," he said grinning; "some son of a bitch might shoot me." Perhaps he declined because of the possibility of being shot accidentally. But the coal fields in West Virginia were seething at that time with anger towards mining companies and their officers. Granny Pritt spoke of miners' families being thrown out of company houses during strikes:

> The company would just come along with its bullies 'en scabs, throw everything we had out onto the county road, in the mud 'en snow, women 'en babies, 'en sick ones, wouldn't give a body a chance to grab a coat

If there was labor union activity at Spruce Knob it was conducted quietly. Father and mother never spoke of it. Mother was a supporter of the miner's labor union—certainly more openly before she married my father and after he died.

I suspect Daddy sympathized with the union, and hoped that he would never have to choose sides between the company he worked for and the men he supervised. But, I never really knew how he felt about unions. I know how I feel. I once took part in trying, unsuccessfully, to start a union in a woolen mill where I worked for a time in the late 1930's

With Joe Postek's help, we linked the tipple on the original grade with the monitors' track that scars the side of Fork Mountain and drops to the railroad tipple in the bottom land on the bank of Cherry River. Then we followed the logging road around the mountain from the remains of the tipple and the cable car buildings to a depression in the earth overgrown by small trees and blackberry bushes. That was where the new entry was located when Homer Rhodes was superintendent of the Elk Lick coal mines at Spruce Knob.

We walked back on the logging road around Fork Mountain from the new entry, to the cable car house and the tipple and continued on to the main entry. We had left our automobile near the split rock. Together Joe and I had reconstructed many memories of our childhood. And for me many spots of time again floated into my consciousness: the killing of Old Bess, bagging a red squirrel, " catching" and "keeping" an undersized brook trout, discovering a blue crayfish, and standing over Lon Spencer bleeding in the bottom of a coal car. And then, as spots of time do, they linger briefly, fade and disappear—perhaps to reappear in distant tomorrows.

EPILOGUE

SOME OF THE attainments of The Coal Miner's Family and the Rhodes/Guinn/Hanke Tribe from Mooseheart include: seven people with training and skills in a Trade, seven High School Diplomas, three College Degrees, two Master's Degrees, two Commissioned Officers in the United States Navy Reserve, a Doctor of Optometry, a PhD in English and American Literature, and a former Superintendent of Mooseheart,

According to Wikipedia, the Moose International fraternity continues today with nearly 1 million men in roughly 2,000 lodges and nearly 400,000 Women of the Moose in 1,600 lodges in the United States, Canada, Great Britain and Bermuda. Mooseheart has evolved through Social Security and Foster Parent Programs as a "Child City & School" with an enrollment of about 300 "children and teens in need." Moosehaven, Jacksonville, FL, thrives as a home for its aging brotherhood

For a scholarly treatment of the effect of social development on Orphans Homes see David T. Beito. From Mutual Aid to the Welfare State, the University Press of North Carolina, 2000. Beito devotes Chapter 4, pp 63-80 to "The Child City." It is the best and fullest account of the Mooseheart story that I have found. Beito points out on that the expense per child of caring for one student was $800 in 1929. He concludes in the final sentence of his study based interviews with alumni: "The financial cost of Mooseheart may have been high, but it was money well spent."

THE END

www.ingramcontent.com/pod-product-compliance
Lightning Source LLC
Chambersburg PA
CBHW022245290526
45785CB00015B/248

* 9 7 8 1 4 5 2 0 0 9 5 0 6 *